Protein Lipidation Protocols

METHODS IN MOLECULAR BIOLOGY™

John M. Walker, Series Editor

133. **Gene Targeting Protocols,** edited by *Eric B. Kmiec, 1999*
132. **Bioinformatics Methods and Protocols,** edited by *Stephen Misener and Stephen A. Krawetz, 1999*
131. **Flavoprotein Protocols,** edited by *S. K. Chapman and G. A. Reid, 1999*
130. **Transcription Factor Protocols,** edited by *Martin J. Tymms, 1999*
129. **Integrin Protocols,** edited by *Anthony Howlett, 1999*
128. **NMDA Protocols,** edited by *Min Li, 1999*
127. **Molecular Methods in Developmental Biology:** Xenopus *and Zebrafish*, edited by *Matt Guille, 1999*
126. **Developmental Biology Protocols, Volume II,** edited by *Rocky S. Tuan and Cecilia W. Lo, 1999*
125. **Developmental Biology Protocols, Volume I,** edited by *Rocky S. Tuan and Cecilia W. Lo, 1999*
124. **Protein Kinase Protocols,** edited by *Alastair Reith, 1999*
123. ***In Situ* Hybridization Protocols (2nd ed.),** edited by *Ian A. Darby, 1999*
122. **Confocal Microscopy Methods and Protocols,** edited by *Stephen W. Paddock, 1999*
121. **Natural Killer Cell Protocols:** *Cellular and Molecular Methods,* edited by *Kerry S. Campbell and Marco Colonna, 1999*
120. **Eicosanoid Protocols,** edited by *Elias A. Lianos, 1999*
119. **Chromatin Protocols,** edited by *Peter B. Becker, 1999*
118. **RNA–Protein Interaction Protocols,** edited by *Susan R. Haynes, 1999*
117. **Electron Microscopy Methods and Protocols,** edited by *Nasser Hajibagheri, 1999*
116. **Protein Lipidation Protocols,** edited by *Michael H. Gelb, 1999*
115. **Immunocytochemical Methods and Protocols (2nd ed.),** edited by *Lorette C. Javois, 1999*
114. **Calcium Signaling Protocols,** edited by *David Lambert, 1999*
113. **DNA Repair Protocols:** *Eukaryotic Systems,* edited by *Daryl S. Henderson, 1999*
112. **2-D Proteome Analysis Protocols,** edited by *Andrew J. Link 1999*
111. **Plant Cell Culture Protocols,** edited by *Robert Hall, 1999*
110. **Lipoprotein Protocols,** edited by *Jose M. Ordovas, 1998*
109. **Lipase and Phospholipase Protocols,** edited by *Mark H. Doolittle and Karen Reue, 1999*
108. **Free Radical and Antioxidant Protocols,** edited by *Donald Armstrong, 1998*
107. **Cytochrome P450 Protocols,** edited by *Ian R. Phillips and Elizabeth A. Shephard, 1998*
106. **Receptor Binding Techniques,** edited by *Mary Keen, 1998*
105. **Phospholipid Signaling Protocols,** edited by *Ian M. Bird, 1998*
104. **Mycoplasma Protocols,** edited by *Roger J. Miles and Robin A. J. Nicholas, 1998*
103. ***Pichia* Protocols,** edited by *David R. Higgins and James M. Cregg, 1998*
102. **Bioluminescence Methods and Protocols,** edited by *Robert A. LaRossa, 1998*
101. **Mycobacteria Protocols,** edited by *Tanya Parish and Neil G. Stoker, 1998*
100. **Nitric Oxide Protocols,** edited by *Michael A. Titheradge, 1998*
99. **Human Cytokines and Cytokine Receptors,** edited by *Reno Debets and Huub Savelkoul, 1999*
98. **Forensic DNA Profiling Protocols,** edited by *Patrick J. Lincoln and James M. Thomson, 1998*
97. **Molecular Embryology:** *Methods and Protocols,* edited by *Paul T. Sharpe and Ivor Mason, 1999*
96. **Adhesion Protein Protocols,** edited by *Elisabetta Dejana and Monica Corada, 1999*
95. **DNA Topoisomerases Protocols:** *II. Enzymology and Drugs,* edited by *Mary-Ann Bjornsti and Neil Osheroff, 1998*
94. **DNA Topoisomerases Protocols:** *I. DNA Topology and Enzymes,* edited by *Mary-Ann Bjornsti and Neil Osheroff, 1998*
93. **Protein Phosphatase Protocols,** edited by *John W. Ludlow, 1998*
92. **PCR in Bioanalysis,** edited by *Stephen J. Meltzer, 1998*
91. **Flow Cytometry Protocols,** edited by *Mark J. Jaroszeski, Richard Heller, and Richard Gilbert, 1998*
90. **Drug–DNA Interaction Protocols,** edited by *Keith R. Fox, 1998*
89. **Retinoid Protocols,** edited by *Christopher Redfern, 1998*
88. **Protein Targeting Protocols,** edited by *Roger A. Clegg, 1998*
87. **Combinatorial Peptide Library Protocols,** edited by *Shmuel Cabilly, 1998*
86. **RNA Isolation and Characterization Protocols,** edited by *Ralph Rapley and David L. Manning, 1998*
85. **Differential Display Methods and Protocols,** edited by *Peng Liang and Arthur B. Pardee, 1997*
84. **Transmembrane Signaling Protocols,** edited by *Dafna Bar-Sagi, 1998*
83. **Receptor Signal Transduction Protocols,** edited by *R. A. John Challiss, 1997*
82. **Arabidopsis Protocols,** edited by *José M Martinez-Zapater and Julio Salinas, 1998*
81. **Plant Virology Protocols:** *From Virus Isolation to Transgenic Resistance*, edited by *Gary D. Foster and Sally Taylor, 1998*
80. **Immunochemical Protocols (2nd. ed.),** edited by *John Pound, 1998*
79. **Polyamine Protocols,** edited by *David M. L. Morgan, 1998*
78. **Antibacterial Peptide Protocols,** edited by *William M. Shafer, 1997*
77. **Protein Synthesis:** *Methods and Protocols,* edited by *Robin Martin, 1998*
76. **Glycoanalysis Protocols (2nd. ed.),** edited by *Elizabeth F. Hounsell, 1998*
75. **Basic Cell Culture Protocols (2nd. ed.),** edited by *Jeffrey W. Pollard and John M. Walker, 1997*
74. **Ribozyme Protocols,** edited by *Philip C. Turner, 1997*
73. **Neuropeptide Protocols,** edited by *G. Brent Irvine and Carvell H. Williams, 1997*
72. **Neurotransmitter Methods,** edited by *Richard C. Rayne, 1997*
71. **PRINS and *In Situ* PCR Protocols,** edited by *John R. Gosden, 1996*
70. **Sequence Data Analysis Guidebook,** edited by *Simon R. Swindell, 1997*

METHODS IN MOLECULAR BIOLOGY™

Protein Lipidation Protocols

Edited by

Michael H. Gelb

University of Washington, Seattle, WA

Humana Press **Totowa, New Jersey**

999 Riverview Drive, Suite 208
Totowa, New Jersey 07512

This publication is printed on acid-free paper. ∞
ANSI Z39.48-1984 (American Standards Institute)
Permanence of Paper for Printed Library Materials.

Cover illustration: Fig. 4(**B**) from Chapter 4, "Imaging Fluorescence Resonance Energy Transfer as Probe of Membrane Organization and Molecular Associations of GPI-Anchored Proteins," by Anne K. Kenworthy and Michael Edidin.

Cover design by Patricia F. Cleary.

For additional copies, pricing for bulk purchases, and/or information about other Humana titles, contact Humana at the above address or at any of the following numbers: Tel.: 973-256-1699; Fax: 973-256-8341; E-mail: humana@humanapr.com; or visit our Website: http://humanapress.com

Printed in the United States of America. 10 9 8 7 6 5 4 3 2 1

Library of Congress Cataloging in Publication Data

Main entry under title:

Methods in molecular biology™.

Protein lipidation protocols / edited by Michael H. Gelb.
p. cm. -- (Methods in molecular biology™ ; v. 116)
Includes bibliographic references and index.
ISBN 0-89603-534-4 (alk. paper)
1. Lipoproteins—Laboratory manuals. 2. Post-translational modification—Laboratory manuals.
I. Gelb, Michael H. II. Series: Methods in Molecular Biology (Totowa, NJ) ; 116.
QP552.L5P77 1999
547'.75—dc21

98-50534
CIP

Preface

It is hard to think of a protein in eukaryotic cells that does not undergo some type of posttranslational modification. The covalent attachment of lipids to proteins, protein lipidation, occurs for a few thousand proteins. Several functions for protein lipidation are known. Protein lipids may target proteins to specific cellular membranes, they may serve as molecular switches that allow cytosol-to-membrane transfer, they may direct protein–protein complexation, and they may stabilize protein structure. In cases such as the fatty acylation of intracellular loops of transmembrane proteins, the functions of the protein lipidations are not well understood.

This volume—*Protein Lipidation Protocols*—provides detailed methodologies for the study of these processes. Since this is a rapidly growing field, many new experimental techniques have been developing over the past few years. All the experimental techniques described in this volume have emerged during this time. The editor has made a special effort to include only those techniques not previously described in a "hands-on" format.

Three areas of protein lipidation are included. The first section deals with glycosyl phosphatidylinositol (GPI)-containing proteins. Protein prenylation is covered next, followed by protein fatty acylation. Very recently, mammalian cell mutants defective in GPI biosynthesis have been obtained, and such cell lines will undoubtedly lead to a more complete understanding of the role of GPI groups in protein function. This work is described in the first two chapters. There is accumulating evidence that GPI groups target proteins to specific sites on cellular membranes such as caveolae, and new methods for visualizing the location of GPI-anchored proteins on cell membranes are described in the next four chapters. Finally, Chapters 7 and 8 describe the very recent addition of techniques for determining the chemical structure of GPI-anchors.

The middle portion of this volume is concerned with protein prenylation, the attachment of 15-carbon farnesyl and 2-carbon geranylgeranyl groups to proteins. Interest in protein prenylation has escalated in the last five years because of its medicinal impact. The cancer-causing protein Ras is farnesylated, and this lipid is required for the ability of Ras to switch cells into a proliferating mode. Approximately 30% of human tumor cell lines contain activated Ras proteins, and there is a massive, worldwide effort to develop inhibitors of the protein farnesyltransferase that attaches farnesyl groups to proteins, including Ras. Chapters 9 and 10 describe a novel technique for radiolabeling the prenyl groups of proteins in eukaryotic cells. Chapter 11 describes a recent development in the manipulation of yeast protein farnesyltransferase. Chapter 12 describes a whole-cell system for studying the functions of protein prenyl groups. Finally, Chapter 13 describes ways to quantify the binding of prenylated peptides to membranes.

The last section of the book is focused on protein fatty acylation. Chapter 14 describes the assay and purification of the mammalian enzyme that attaches myristoyl groups to the N-terminus of specific proteins. Although the yeast protein myristoyltranfserase has been purified, the mammalian homolog has been difficult to obtain. The volume ends with a chapter on an exciting lysosomal enzyme that cleaves fatty acyl groups from proteins. This enzyme is the first such protein fatty acylase to have been discovered.

All the chapters of *Protein Lipidation Protocols* contain a brief introduction, followed by detailed methodological descriptions, notes on optimizing the protocols for use in other systems, and some concluding remarks. Each chapter can be read on its own.

The editor is particularly grateful for the cooperation of all the chapter authors. The Series Editor, John Walker, and the production team at Humana Press, headed by Tom Lanigan, have provided invaluable assistance toward the publication of this important volume. The editor is grateful to Ms. Kathlene Bennett for excellent assistance in the process of organizing this volume.

Michael H. Gelb

Contents

Contributors

DOUGLAS A. ANDRES • *Department of Biochemistry, University of Kentucky, Lexington, KY*

ELISABETH CANIVENC-GANSEL • *Institut De Biochimie, Universite De Fribourg, Switzerland*

ANDREAS CONZELMANN • *Institut De Biochimie, Universite De Fribourg, Switzerland*

ALBERTO CORSINI • *Howard Hughes Medical Institute, University of Washington, Seattle, WA*

DEAN C. CRICK • *Department of Biochemistry, University of Kentucky, Lexington, KY*

THOMAS DUDLER • *Departments of Chemistry and Biochemistry, University of Washington, Seattle, WA*

MICHAEL EDIDIN • *Department of Biology, Johns Hopkins University, Baltimore, MD*

CHRISTOPHER C. FARNSWORTH • *Howard Hughes Medical Institute, University of Washington, Seattle, WA*

MICHAEL A. J. FERGUSON • *Department of Biochemistry, University of Dundee, Dundee, Scotland*

BRIAN S. FINLIN • *Department of Biochemistry, University of Kentucky, Lexington, KY*

MICHAEL H. GELB • *Departments of Chemistry and Biochemistry, University of Washington, Seattle, WA*

JOHN A. GLOMSET • *Howard Hughes Medical Institute, University of Washington, Seattle, WA*

SANDRA L. HOFMANN • *Department of Medicine and Hamon Center for Therapeutic Oncology Research, University of Texas Southwestern Medical Center, Dallas, TX*

ANNE K. KENWORTHY • *Department of Biology, Johns Hopkins University, Baltimore, MD*

MICHAEL P. LISANTI • *Department of Molecular Pharmacology and The Albert Einstein Cancer Center, Albert Einstein College of Medicine, Bronx, NY*

JUN LIU • *Department of Pathology, Beth Israel Hospital, Harvard Medical School, Boston, MA*

JUI-YUN LU • *Department of Medicine and Hamon Center for Therapeutic Oncology Research, University of Texas Southwestern Medical Center, Dallas, TX*

SATYAJIT MAYOR • *The National Centre of Biological Sciences, TIFR Centre, Bangalore, Karnataka, India*

PAUL MCGEADY • *Howard Hughes Medical Institute, University of Washington, Seattle, WA*

RAJALA V. S. RAJU • *Department of Pathology and Saskatoon Cancer Center, College of Medicine, University of Saskatchewan, Saskatoon, Saskatchewan, Canada*

FULVIO REGGIORI • *Institut De Biochimie, Universite De Fribourg, Switzerland*

MASSIMO SARGIACOMO • *Department of Cell Biology, Albert Einstein College of Medicine, Bronx, NY*

PHILIPP E. SCHERER • *Department of Cell Biology, Albert Einstein College of Medicine, Bronx, NY*

JAN E. SCHNITZER • *Department of Pathology, Beth Israel Hospital, Harvard Medical School, Boston, MA*

RAJENDRA K. SHARMA • *Department of Pathology and Saskatoon Cancer Center, College of Medicine, University of Saskatchewan, Saskatoon, Saskatchewan, Canada*

JOHN R. SILVIUS • *Department of Biochemistry, McGill University, Montreal, Quebec, Canada*

VICTORIA L. STEVENS • *Emory University School of Medicine, Atlanta, GA*

FUYUHIKO TAMANOI • *Department of Microbiology and Molecular Genetics, UCLA, Los Angeles, CA*

JUN URANO • *Department of Microbiology and Molecular Genetics, UCLA, Los Angeles, CA*

LINDA A. VERKRUYSE • *Hamon Center for Therapeutic Oncology Research, University of Texas Southwestern Medical Center, Dallas, TX*

CHARLES J. WAECHTER • *Department of Biochemistry, University of Kentucky, Lexington, KY*

NICOLE ZITZMANN • *Department of Biochemistry, Oxford Glycobiology Institute, Oxford University, Oxford, UK*

1

In Vitro Analysis of GPI Biosynthesis in Mammalian Cells

Victoria L. Stevens

1. Introduction

1.1. Background

The basic strategy used in most assays of activities involved in the biosynthesis of glycosylphosphatidylinositol (GPI) in mammalian cells is the same as is employed for other lipid biosynthetic pathways. That is, radioactivity is transferred from a water-soluble substrate into a lipophilic product. After the reaction is complete, the differential solubility of the substrate and product(s) is exploited to separate these radiolabeled compounds. In GPI biosynthesis, at least one of the substrates in each step and all of the enzymes in the pathway are membrane-associated and localized to the endoplasmic reticulum. Therefore, multiple GPI biosynthetic activities, as well as some of the substrates for later steps in the pathway, are present in the cellular preparations used in the assays. For this reason, multiple intermediates in GPI biosynthesis are usually generated in a single reaction. Although it is possible to optimize the assay conditions for one step, it is usually impossible to study one reaction independently with this type of cell-free system.

Assays for individual reactions in GPI biosynthesis are possible if synthetic GPI intermediates are available. To date, the second and third reactions in the pathway have been measured in mammalian cells with exogenously supplied GlcNAc-PI *(1)* and GlcN-PI *(2)*, respectively. In the latter case, a short-chain (dioctanoyl) analog of the GPI intermediate was used. Because the short-chain analogs are more water-soluble, they are much easier to deliver to the membranes used as a source of the GPI biosynthetic activities.

From: *Methods in Molecular Biology, Vol. 116: Protein Lipidation Protocols*
Edited by: M. H. Gelb © Humana Press Inc., Totowa, NJ

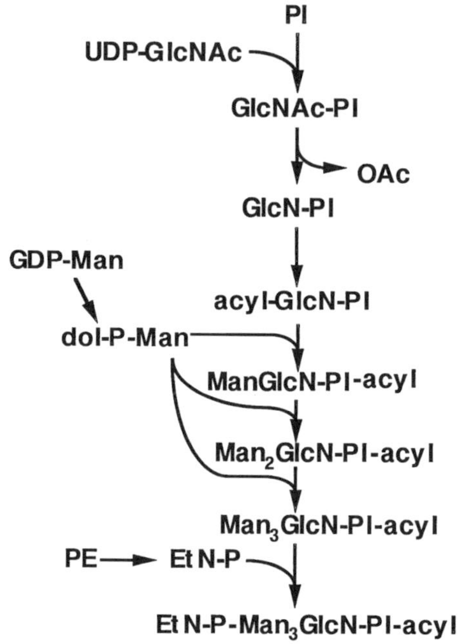

Fig. 1. GPI biosynthesis in mammals and yeast.

1.1.1. Pathway for GPI Biosynthesis in Mammalian Cells

The biosynthesis of GPI proceeds by the sequential addition of carbohydrates to phosphatidylinositol (PI), as is shown in **Fig. 1** (reviewed in **refs. *3*** and ***4***) . In the first step, N-acetylglucosamine (GlcNAc) is transferred from UDP-GlcNAc to PI ***(5,6)***. The resulting product, GlcNAc-PI, is then deacetylated to glucosamine-PI (GlcN-PI) in the second step. Next, in a reaction that only occurs at this step in mammals and yeast, an acyl chain is added to the inositol ring to form GlcN-PI(acyl) ***(7,8)***. Mannoses are then sequentially transferred to the growing GPI core. The endogenous source of all three mannoses is dolichol-phospho-mannose ***(9)***, which is made from GDP-mannose and dolichol-phosphate. Finally, a phosphoethanolamine residue is transferred from phosphatidylethanolamine to the third mannose to complete the GPI core ***(10,11)***. Analysis of the structures of GPI precursors from normal and Thy-1-deficient murine lymphoma cell lines suggests that there may be one or two extra phosphoethanolamines added to the first and second mannoses before addition of the final phosphoethanolamine ***(12–16)***. However, the exact sequence of steps leading to these precursors, and whether they are really intermediates in the synthesis of the GPI anchor in all cases, is not known.

All the intermediates in GPI biosynthesis should be detected if UDP-[6-^{3}H]GlcNAc is used in the assay. However, it is really only practical to use this radiolabeled sugar nucleotide to assay the first three steps in the pathway. Conditions to measure at least the first two mannose addition reactions using GDP-[2-^{3}H]mannose have been described ***(17)***. To date, no in vitro assays for the addition of the third mannose, the terminal phosphoethanolamine, or the extra phosphoethanolamines which extend from the GPI core, have been developed.

1.2. Sources of GPI Biosynthetic Enzymes

The enzymatic activities necessary for GPI biosynthesis are localized to the endoplasmic reticulum ***(18)***. Therefore, cell lysates, permeabilized cells, microsomal preparations, and isolated endoplasmic reticulum all will contain these enzymes and can be used for the assays described here. All of these preparations also contain phosphatidylinositol in sufficient quantities so that detection of the initial GPI intermediates upon labeling with UDP-[6-^{3}H]GlcNAc should be possible. However, the levels of later intermediates in the pathway are much lower in any of these membranes, which may explain why detection of intermediates with GDP-[2-^{3}H]mannose is so difficult.

2. Materials

2.1. Cell Lysis

1. Phosphate buffered saline (PBS).
2. Lysis buffer: 10 m*M* HEPES, pH 7.5, 1 µg/mL leupeptin, 0.1 m*M* N$^\alpha$-tosyl-L-lysine chloromethyl ketone (TLCK). Add fresh protease inhibitors to cold lysis buffer.
3. Bath sonicator.

2.2. Permeabilization of Cells

1. Streptolysin O (Gibco BRL).
2. Dithiothreitol (DTT): supplied in the 10X activating solution from Gibco BRL.
3. PBS (Ca^{2+}- and Mg^{2+}-free).
4. Lysis buffer: 10 m*M* HEPES, pH 7.5, 1 µg/mL leupeptin, 0.1 m*M* N$^\alpha$-TLCK.

2.3. Cellular Fractionation

1. PBS.
2. Fractionation buffer: 0.25 *M* sucrose, 0.5 m*M* DTT, 0.1 m*M* TLCK, 1 µg/mL leupeptin.
3. Cell disruption bomb, nitrogen gas.
4. High-speed centrifuge.
5. Ultracentrifuge.
6. Microsome buffer: 10 m*M* HEPES, pH 7.5, 0.5 m*M* DTT, 0.1 m*M* TLCK, 1 µg/mL leupeptin.

7. Sucrose solutions of 38, 30, and 20% sucrose in 10 m*M* HEPES, pH 7.5, 1 m*M* DTT.
8. Glycerol.
9. Swing bucket rotor.

2.4. Labeling with UDP-[6-^{3}H]GlcNAc

1. Incubation buffer: 60 m*M* HEPES, pH 7.5, 30 m*M* $MgCl_2$, 3 m*M* DTT, 0.6 μg/mL leupeptin, 1.2 μ*M* tunicamycin.
2. 50 m*M* ATP.
3. 50 m*M* GTP.
4. UDP-[6-^{3}H]GlcNAc (5–15 Ci/mmol, American Radiolabeled Chemicals, St. Louis, MO).
5. 50 m*M* dimercaptopropanol.
6. 50 m*M* EDTA.
7. Water bath at 37°C.
8. 13 × 100 glass screw top tubes with teflon-coated caps.
9. Chloroform–methanol–0.1 *M* HCl, 1:2:0.5 (v/v).

2.5. Labeling with GDP-[2-^{3}H]mannose

1. Incubation buffer (-tunicamycin): 60 m*M* HEPES, pH 7.5, 30 m*M* $MgCl_2$, 3 m*M* DTT, 0.6 μg/mL leupeptin.
2. 50 m*M* ATP.
3. 50 m*M* GTP.
4. UDP-GlcNAc.
5. GDP-[1-^{3}H]mannose (5–15 Ci/mmol, American Radiolabeled Chemicals).
6. 50 m*M* dimercaptopropanol.
7. 50 m*M* EDTA.
7. Water bath at 37°C.
8. 13 × 100 glass screw-top tubes with teflon-coated caps.
9. Chloroform–methanol, 1:1 (v/v).

2.6. Synthesis and Purification of [6-^{3}H]GlcNAc-PI

1. Chloroform–methanol–H_2O, 2:3:1 (v/v).
2. UDP-[6-^{3}H]GlcNAc.
3. 50 m*M* ammonium acetate.
4. Pasteur pipet.
5. Glass wool.
6. DEAE cellulose pre-equilibrated with chloroform–methanol–H_2O, 2:3:1 (v/v).
7. Chloroform–methanol–50 m*M* ammonium acetate, 2:3:1 (v/v).
8. Speed-Vac concentrator.
9. Scintillation vials.
10. Ethanol.

2.7. Extraction of [6-^{3}H]GlcNAc-Labeled Products

1. Chloroform.
2. H_2O.

3. Tabletop centrifuge.
4. Pre-equilibrated acidic upper phase: Prepare by mixing chloroform–methanol–0.1 *M* HCl, 2:2:1.5 (v/v), in a separatory funnel. Let layers separate completely. Collect upper phase.
5. Speed-Vac concentrator.

2.8. Extraction of [1-^{3}H]-Mannose-Labeled Products

1. Tabletop centrifuge.
2. Chloroform–methanol–H_2O, 1:1:0.3 (v/v).
3. Speed-Vac concentrator.
4. H_2O-saturated butanol.
5. H_2O.

2.9. Thin Layer Chromatography (TLC) of Products

1. TLC tank.
2. Silica gel 60 (20 × 20 cm) TLC plates (E. Merck, VWR Scientific, Atlanta, GA).
3. Chloroform–methanol–1 *M* ammonium hydroxide, 10:10:3 (v/v).
4. Imaging scanner capable of detecting ^{3}H or En^{3}Hance spray (NEN/Dupont) and Kodak XAR-5 film.

3. Methods

These methods have been developed for use with cultured cells. In some cases, the procedure may have to be modified slightly to optimize conditions for different types of cells or tissues.

3.1. Preparation of Membranes for Analysis of GPI Biosynthesis

Each of these methods will yield preparations that can be used in each of the assays described in **Subheading 3.2.**

3.1.1. Cell Lysates

1. Wash cells with PBS by centrifugation (5 min at 800*g*).
2. Resuspend the cells in lysis buffer at a density of approximately 1.2×10^8 cells/mL.
3. Disrupt cells by three cycles of sonic irradiation (10 s each).

3.1.2. Permeabilized Cells

1. Solubilize the streptolysin O by adding distilled water to generate a stock solution of 1000 U/mL.
2. Activate as much of the stock solution as needed by incubating the streptolysin O with 2 m*M* DTT for 15 min at 37°C. If using Streptolysin O obtained from Gibco-BRL, this activation is accomplished by adding one part of the 10X activating solution per nine parts of the streptolysin O stock solution.
3. Wash cells twice with PBS by centrifugation (5 min at 800*g*).
4. Resuspend cells in cold streptolysin O solution at a density of 50–100 U/10^7 cells. Incubate on ice for 20 min to allow the toxin to insert into the membrane.

5. Pellet the cells by centrifugation (5 min at 800*g* at 4°C). Wash cells once with cold PBS.
6. Resuspend in lysis buffer at a concentration of approx 10^8 cells/mL.

3.1.3. Microsomes

1. Wash cells twice with PBS by centrifugation (5 min at 800*g*).
2. Resuspend the cells in fractionation buffer at a density of 0.5–1 × 10^8 cells/mL.
3. Lyse the cells by nitrogen cavitation using 450 psi for 15–30 min.
4. Centrifuge at 10,000*g* for 5 min to remove unbroken cells and nuclei.
5. Centrifuge the resulting supernatant (18,000*g*, 15 min) to remove mitochondria.
6. Centrifuge the supernatant at 100,000*g* for 1 h, to pellet the microsomes.
7. Resuspend this pellet in microsome buffer. Recentrifuge at 100,000*g* for 1 h, to wash the microsomes.
8. Resuspend the final microsomal pellet microsome buffer containing 10% glycerol at a protein concentration of approx 70 mg/mL.

3.1.4. Endoplasmic Reticulum

1. Wash cells twice with PBS by centrifugation (5 min at 800*g*).
2. Resuspend the cells in fractionation buffer at a density of 0.5–1 × 10^8 cells/mL.
3. Lyse the cells by nitrogen cavitation using 450 psi for 15–30 min.
4. Centrifuge at 10,000*g* for 15 min at 4°C, to pellet unbroken cells and nuclei.
5. Layer the 4.06 mL of the resulting postnuclear supernatant onto a preformed sucrose gradient consisting of 2.52 mL 38% sucrose, 1.26 mL 30% sucrose, and 1.26 mL 20% sucrose.
6. Centrifuge this gradient 2 h at 28,000*g* in a Sorvall TH-641 rotor.
7. Collect four fractions of 1.96 (1), 2.1 (2), 2.38 (3), and 2.66 (4) mL from the top of the tube. Resuspend the pellet in 1 mL of microsome buffer to make fraction 5. The endoplasmic reticulum will be enriched in fractions 4 and 5.

3.2. In Vitro Biosynthesis of GPI Intermediates from Radiolabeled Precursors

At least the first five steps in GPI biosynthesis can be assayed in vitro with various membrane preparations. The choice of radiolabeled precursor for the assay will depend on which reaction or reactions the investigator wants to measure.

3.2.1. UDP-N-Acetylglucosamine

1. Mix incubation components in a 13 × 100 glass screw-top tube in a total volume of 300 µL. Components should include 50 to 100 µL of the appropriate membrane preparation (about 300 µg protein, measured using the bicinchoninic acid assay of Smith *[19]*), 50 µL incubation buffer, and 1 m*M* ATP. The following effectors should be added to optimize synthesis of various intermediates: for GlcNAc-PI, no additions; for GlcN-PI, 0.1–1 m*M* GTP; for GlcN-PI(acyl),

0.1–1 m*M* GTP and 1 μ*M* CoA; and mannose-containing intermediates, 0.1–1 m*M* GTP, 1 μ*M* CoA, and 200 μ*M* GDP-mannose.

2. Initiate reaction by adding 1 μCi UDP-[6-^{3}H]GlcNAc.
3. Incubate at 37°C for 1 h.
4. Stop reaction by the addition of 3.5 mL chloroform–methanol–0.1 *M* HCl, 1:2:0.5 (v/v).

3.2.2. GDP-Mannose

1. Mix incubation components in a 13 × 100 glass screw-top tube in a total volume of 300 μL. Components should include 50–100 μL of the appropriate membrane preparation (about 300 μg of protein), 50 μL of incubation buffer (without tunicamycin), 1 m*M* ATP, 1 m*M* GTP, 1 μ*M* CoA, and 100 μ*M* UDP-GlcNAc.
2. Initiate reaction by adding 1 μCi GDP-[1-^{3}H]mannose.
3. Incubate at 37°C for 1 h.
4. Stop reaction by the addition of 2 mL of chloroform–methanol, 1:1 (v/v).

3.2.3. N-Acetylglucosamine-Phosphatidylinositol

1. Synthesize this substrate from UDP-[6-^{3}H]GlcNAc, as described in **Subheading 3.2.1.**, with the following changes: Extend the 37°C incubation time to 2 h, increase the amount of UDP-[6-^{3}H]GlcNAc in the reaction to 4 μCi, double the incubation volume and microsomal protein concentration, and add only 1 m*M* ATP to the reaction. Extract the products as described below in **Subheading 3.3.1.**, and dissolve them in 100 μL of chloroform–methanol–H_2O, 2:3:1, (v/v).
2. Prepare a column in a Pasteur pipet by plugging the narrow portion with glass wool and adding approx 1 mL DEAE cellulose pre-equilibrated with chloroform–methanol–H_2O, 2:3:1, (v/v). Apply products to the column.
3. Remove the uncharged GPI intermediates (GlcN-PI and GlcN-PI[acyl]) by washing the column with 4 column volumes of chloroform–methanol–H_2O, 2:3:1, (v/v).
4. Elute the [^{3}H]GlcNAc-PI from the column with 3–4 mL of chloroform–methanol–50 m*M* ammonium acetate, 2:3:1, (v/v). Collect 1-mL fractions in 13 × 100 glass tubes.
5. Determine which fractions contain [^{3}H]GlcNAc-PI by scintillation counting of a 10-μL aliquot of each. Pool fractions containing this GPI intermediate, and add chloroform and H_2O to bring the chloroform–methanol–H_2O proportions to 2:2:1.8 (v/v). Centrifuge the solution to separate phases (5 min at 1200*g*).
6. Remove the upper phase by aspiration and dry the lower under vacuum. Dissolve the [^{3}H]GlcNAc-PI in ethanol at a concentration of 10000 cpm/2 μL.
7. For synthesis of GPI intermediates from [^{3}H]GlcNAc-PI, mix incubation components in a 13 × 100 glass screw-top tube in a total volume of 300 μL. Components should include 50–100 μL of the appropriate membrane preparation (about 300 μg of protein) and 50 μL of incubation buffer (without tunicamycin). Include 1 m*M* GTP, 1 μ*M* CoA, and/or 200 μ*M* GDP-mannose, depending on the desired products.
8. Initiate reaction by adding 10,000 cpm of [6-^{3}H]GlcNAc-PI (in 2 μL ethanol).
9. Incubate at 37°C for 1 h.

10. Stop reaction by the addition of 3.5 mL of chloroform/methanol/0.1 *M* HCl, 1:2:0.5 (v/v).

3.3. Extraction of Radiolabeled Products

The first three intermediates in GPI biosynthesis are completely chloroform-soluble, and therefore can be extracted using the method of Bligh and Dyer *(20)*. The mannose-containing intermediates are more hydrophilic. Therefore, the extraction solvent used must be less hydrophobic, for quantitative recovery of these products.

3.3.1. Chloroform-Soluble Products

1. Extract the first three intermediates in GPI biosynthesis from the UDP-[6-^{3}H]GlcNAc and [6-^{3}H]GlcNAc-PI incubation by adding 1 mL chloroform and 1 mL H_2O to the 13 × 100 tube containing the reaction products generated in **Subheading 3.2.2., step 4** or **Subheading 3.2.3., step 10** (in 3.5 mL chloroform–methanol–0.1 N HCl, 1:2:0.5 [v/v]).
2. Mix by vortexing. Separate into two phases by centrifugation (5 min at 1200*g*).
3. Remove the top phase by aspiration. Wash the lower phase once with 3 mL of pre-equilibrate acidic upper phase by centrifugation (5 min at 1200*g*).
4. Remove the upper phase by aspiration. Dry the lower phase under vacuum in a Speed-Vac concentrator.

3.3.2. Mannose-Containing Products

1. Extract mannose-containing GPI intermediates by centrifuging the products obtained in incubations with GDP-mannose (*see* **Subheading 3.2.2.**, which are in chloroform–methanol, 1:1 [v/v]) at 1200*g* for 5 min.
2. Transfer the lipid-containing supernatant to a fresh 13 × 100 glass screw-top tube.
3. Wash the pellet once with 2 mL chloroform–methanol–H_2O, 1:1:0.3 (v/v) by centrifugation (1200*g* for 5 min). Combine the resulting supernatant with the first and dry under vacuum with a Speed-Vac concentrator.
4. Resuspend the lipids in 400 µL H_2O-saturated butanol. Transfer to a 1.5-mL microcentrifuge tube.
5. Add 400 µL H_2O and mix by vortexing. Separate phases by centrifugation in a microcentrifuge for 3 min.
6. Remove the water phase, and wash the butanol once with 400 µL H_2O-saturated butanol by centrifugation (3 min in microcentrifuge).
7. Collect final butanol phase and dry under vacuum with a Speed-Vac concentrator.

3.4. Analysis of Products

1. Dissolve radiolabeled GPI intermediates in 10–20 µL chloroform–methanol, 1:1 (v/v).
2. Spot on silica gel 60 TLC plates.
3. Place TLC plate in tank pre-equilibrated with solvent by lining with paper with 150–200 mL chloroform–methanol–1 *M* ammonium hydroxide, 10:10:3 (v/v). Remove from tank when solvent is within 1 cm of the top of the TLC plate.

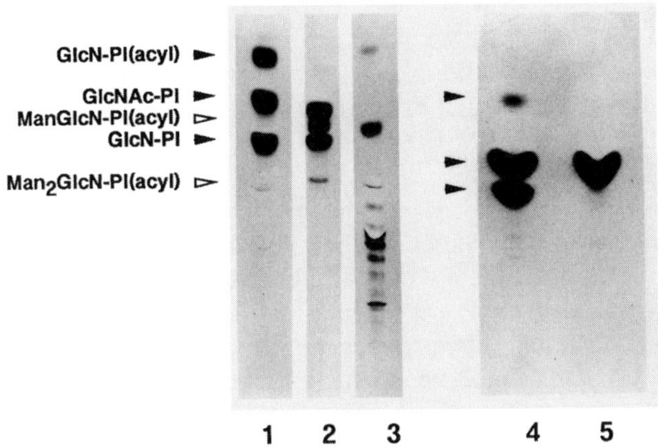

Fig. 2. GPI intermediates synthesized in vitro. Microsomes isolated from wild-type EL4 cells (lanes 1–3) or alkaline phosphatase-transfected G9PLAP (lane 4) or G9PLAP.85 (lane 5) cells were incubated UDP-[6-^{3}H]GlcNAc (lanes 1 and 2), GDP-[2-^{3}H]mannose (lane 3), or [6-^{3}H]GlcNAc-PI (lanes 4 and 5), as described in **Subheading 3.2.** Additions to the incubations were: lane 1, ATP, GTP, and CoA (to optimize synthesis of the third product); lane 2, ATP, GTP, CoA, and GDP-mannose (to allow synthesis of ManGlcN-PI[acyl] and Man$_2$GlcN-PI[acyl]); lane 3, ATP, GTP, CoA, and UDP-GlcNAc (to optimize synthesis of ManGlcN-PI[acyl] and Man$_2$GlcN-PI[acyl]); lanes 4 and 5, ATP and GTP. No GlcN-PI or GlcN-PI(acyl) is seen in lane 5, because the G9PLAP.85 cell line is deficient in the GlcNAc-PI deacetylase activity *(21)*.

4. Air dry plate completely.
5. Scan plate with a imaging scanner capable of detecting ^{3}H or spray plate with En^{3}Hance, air-dry, and expose to Kodak XAR-5 film for 3–5 d to visualize the radiolabeled GPI precursors.
6. The products generated in a typical assay using microsomes isolated from the murine lymphoma cell line EL4 (lanes 1–3), the CHO-derived alkaline phosphatase-transfected G9PLAP (lane 4), or G9PLAP.85 (lane 5, described in **ref.** ***21***) are shown in **Fig. 2**. The radiolabeled substrate used were UDP-[6-^{3}H]GlcNAc (lanes 1 and 2), GDP-[2-^{3}H]Man (lane 3), or [6-^{3}H]GlcNAc-PI (lanes 4 and 5).

4. Notes

1. Cell lysates and permeabilized cells must be used immediately after preparation. Microsomes and endoplasmic reticulum preparations can be stored at –80°C for up to 6 mo before use.
2. Marker enzymes should be measured in the fractions recovered from the sucrose gradient to determine the distribution of the various cellular membranes. Typical marker enzymes used for this purpose are alkaline phosphodiesterase I (plasma membrane *[22]*), α-mannosidase II (Golgi *[22]*), and dolichol-phospho-mannose synthase (endoplasmic reticulum *[18]*).

3. The addition of ATP to incubations in which GPI intermediates are being synthesized from UDP[6-^{3}H]GlcNAc has been shown to inhibit the degradation of this sugar nucleotide by non-GPI related activities *(1)*. With membranes derived from some cell lines, the addition of either 1 m*M* dimercaptopropanol (murine lymphoma cells, *[1]*) or 1 mM EDTA (CHO *[21]*) has been shown to assist in the inhibition of these activities.
4. Pre-equilibration of DEAE cellulose with chloroform–methanol–H_2O, 2:3:1 (v/v) is a stepwise procedure, in which the hydrophobicity of this ion exchanger is gradually increased, using a series of solutions. The details of this procedure are described by Radika and Raetz *(23)*.
5. If mannose-containing intermediates are synthesized from UDP[6-^{3}H]GlcNAc, the protocol should be modified to maximize the extraction of these more hydrophilic products. Therefore, the reaction should be stopped with the addition of 2 mL chloroform–methanol, 1:1 (v/v), and the method described in **Subheading 3.2.2.** for the extraction of more amphipathic products followed.
6. The presence of detergents in any of the assays described above disrupts GPI biosynthesis, because of surface dilution of the substrates from the enzymes. Some of the enzymes may also be inhibited by certain detergents. Therefore, detergents should be avoided in these assays, and, if one is required to deliver some hydrophobic molecule to the membrane, the levels should be kept below 0.01%.

References

1. Stevens, V. L. (1993) Regulation of glycosylphosphatidylinositol biosynthesis by GTP. Stimulation of N-acetylglucosamine-phosphatidylinositol deacetylation. *J. Biol. Chem.* **268,** 9718–9724.
2. Doerrler, W. T., Ye, J., Falck, J. R., and Lehrman, M. A. (1996) Acylation of glucosaminyl phosphatidylinositol revisited. Palmitoyl-CoA dependent palmitoylation of the inositol residue of a synthetic dioctanoyl glucosaminyl phosphatidylinositol by hamster membranes permits efficient mannosylation of the glucosamine residue. *J. Biol. Chem.* **271,** 27,031–27,038.
3. McConville, M. J. and Ferguson, M. A. J. (1993) The structure, biosynthesis and function of glycosylated phosphatidylinositols in the parasitic protozoa and higher eukaryotes. *Biochem. J.* **294,** 305–324.
4. Stevens, V. L. (1995) Biosynthesis of glycosylphosphatidylinositol membrane anchors. *Biochem. J.* **310,** 361–370.
5. Masterson, W. J., Doering, T. L., Hart, G. W., and Englund, P. T. (1989) A novel pathway for glycan assembly: biosynthesis of the glycosyl-phosphatidylinositol anchor of the trypanosome variant glycoprotein. *Cell* **56,** 793–800.
6. Doering, T. L., Masterson, W. J., Englund, P. T., and Hart, G. W. (1989) Biosynthesis of the glycosyl phosphatidylinositol membrane anchor of the trypanosome variant surface glycoprotein. *J. Biol. Chem.* **264,** 11,168–11,173.

7. Costello, L. C. and Orlean, P. (1992) Inositol acylation of a potential glycosyl phosphatidyl-inositol anchor precursor from yeast requires acyl coenzyme A. *J. Biol. Chem.* **267,** 8599–8603.
8. Urakaze, M., Kamitani, T., DeGasperi, R., Sugiyama, E., Chang, H., and Yeh, E. T. H. (1992) Identification of a missing link in glycosylphosphatidylinositol anchor biosynthesis in mammalian cells. *J. Biol. Chem.* **267,** 6459–6462.
9. Menon, A. K., Mayor, S., and Schwarz, R. T. (1990) Biosynthesis of glycosyl-phosphatidyl-inositol lipids in *Trypanosoma brucei*: involvement of mannosyl-phosphoryldolichol as the mannose donor. *EMBO J.* **9,** 4249–4258.
10. Menon, A. K. and Stevens, V. L. (1992) Phosphatidylethanolamine is the donor of the ethanolamine residue linking a glycosylphosphatidylinositol anchor to protein. *J. Biol. Chem.* **267,** 15,277–15,280.
11. Menon, A. K., Eppinger, M., Mayor, S., and Schwarz, R. T. (1993) Phosphatidylethanolamine is the donor of the terminal phosphoethanolamine group in trypanosome glycosylphospha-tidylinositols. *EMBO J.* **12,** 1907–1914.
12. Puoti, A., Desponds, C., Fankhauser, C., and Conzelmann, A. (1991) Characterization of a glycophospholipid intermediate in the biosynthesis of glycosyl-phosphatidylinositol anchors accumulating in the Thy-1-negative lymphoma line SIA$^-$b. *J. Biol. Chem.* **266,** 21,051–21,059.
13. Kamitani, T., Menon, A. K., Hallaq, Y., Warren, C. D., and Yeh, E. T. H. (1992) Complexity of ethanolamine phosphate addition in the biosynthesis of glycosylphosphatidylinositol anchors in mammalian cells. *J. Biol. Chem.* **267,** 24,611–24,619.
14. Hirose, S., Prince, G. M., Sevlever, D., Ravi, L., Rosenberry, T. L., Ueda, E., and Medof, M. E. (1992) Characterization of putative glycoinositol phospholipid anchor precursors in mammalian cells. Localization of phosphoethanolamine. *J. Biol. Chem.* **267,** 16,968–16,974.
15. Puoti, A. and Conzelmann, A. (1993) Characterization of abnormal free glyco-phosphatidyl-inositols accumulating in mutant lymphoma cells of classes B, E, F, and H. *J. Biol. Chem.* **268,** 7215–7224.
16. Puoti, A. and Conzelmann, A. (1992) Structural characterization of free glycolipids which are potential precursors for glycophosphatidylinositol anchors in mouse thymoma cell lines. *J. Biol. Chem.* **267,** 22,673–22,680.
17. Stevens, V. L. and Zhang, H. (1994) Coenzyme A dependence of glycosyl-phosphatidylinositol biosynthesis in a mammalian cell-free system. *J. Biol. Chem.* **269,** 31,397–31,403.
18. Vidugiriene, J. and Menon, A. K. (1993) Early lipid intermediates in glycosyl-phosphatidyl-inositol anchor assembly are synthesized in the ER and located in the cytoplasmic leaflet of the ER membrane bilayer. *J. Cell Biol.* **121,** 987–996.
19. Smith, P. K., Krohn, R. I., Hermanson, G. T., Mallia, A. K., Gartner, F. H., Provenzano, M. D., et al. (1985) *Anal. Biochem.* **150,** 76–85.
20. Bligh, E. A. and Dyer, W. J. (1959) A rapid method of total lipid extraction and purification. *Can. J. Biochem.* **37,** 911–917.

21. Stevens, V. L., Zhang, H., and Harreman, M. (1996) Isolation and characterization of Chinese hamster ovary (CHO) mutant defective in the second step in glycosylphosphatidylinositol biosynthesis. *Biochem. J.* **313,** 253–258.
22. Storrie, B. and Madden, E. A. (1990) Isolation of subcellular organelles. *Methods Enzymol.* **182,** 203–225.
23. Radika, K. and Raetz, C.R.H. (1988) Purification and properties of lipid A disaccharide synthase of *Escherichia coli. J. Biol. Chem.* **263,** 14,859–14,867.

2

Selection of Mammalian Cell Mutants in GPI Biosynthesis

Victoria L. Stevens

1. Introduction

1.1. Background

Mammalian cell mutants defective in glycosylphosphatidylinositol (GPI) biosynthesis have proven very valuable for defining the donors of anchor components ***(1,2)***, understanding the sequence of reactions in the pathway ***(3)***, cloning genes required for the synthesis of this glycolipid ***(4–9)***, and probing the function of the membrane anchor in specific cell types ***(10)***. The isolation of mutants with complete defects in various reactions in this pathway has been possible, because cultured cells suffer no adverse consequences when unable to express GPI-anchored proteins on their surface. Such a defect in a whole organism is lethal ***(11,12)***. To date, cell lines defective in four of the seven reactions required to synthesize GPI in mammalian cells have been isolated and assigned to seven complementation groups. New mutants in GPI will be useful for studying the three reactions for which no defective cells exist, and for continued elucidation of the function of different GPI-anchored proteins in various biological processes.

1.2. Experimental Strategy

Proteins that are normally attached to the external surface of the cell by a GPI anchor are either secreted or retained intracellularly, when the synthesis of this glycolipid is disrupted ***(13–16)***. Therefore, the GPI-anchored proteins normally expressed on the surface of the cell are missing in mutants defective in GPI biosynthesis. This loss of surface expression of a specific GPI-anchored protein(s) has been the experimental basis for selection of mutants in this pathway. As a positive selection, a single mutant cell can be isolated from a large population by repeated depletion of normal cells and enrichment of the cell of interest.

From: *Methods in Molecular Biology, Vol. 116: Protein Lipidation Protocols*
Edited by: M. H. Gelb © Humana Press Inc., Totowa, NJ

Critical parameters that influence the selection of mutant cell lines defective in GPI biosynthesis include the choice of cell line, the mutagen used to introduce random DNA damage to disrupt the gene of interest, and the selection strategy for isolation of the desired mutant cell. The choice of cell line will define the possible GPI-anchored proteins whose loss can be selected for, the ease of applying certain selection strategies, and possibly what step in GPI biosynthesis is likely to be affected by the mutagen. If the purpose of isolating the mutant is to investigate the role of GPI in some cellular process, the possible cell lines that can be used will be limited to those in which this event can be studied. The selection of the mutagen may influence which step in the pathway is defective in the resulting mutants, if the various genes are differentially susceptible to different DNA damaging agents. Finally, the choice of selection strategy may depend on what reagents and equipment is available, but the outcome of the selection should not be influenced by the method chosen.

1.2.1. Choice of Cell Line

Of all the variables involved in the isolation of mammalian cells defective in GPI biosynthesis, the choice of the cell line to select the mutant from is probably the most important one to fully consider before starting this procedure. Numerous considerations should influence this choice, and can be broken down into those that relate to the actual isolation of the mutant, those that pertain to the characterization and utility of the mutant after isolation, and those that may affect the specific mutant that is obtained from the selection.

In considering factors that affect the actual isolation of the mutant in GPI biosynthesis, the cells chosen must express one or more GPI-anchored proteins, which can be used to select cells that have lost this marker from their surface. Although the GPI-anchored protein can either be endogenous to the cell line or introduced by transfection of the appropriate cDNA, it must be expressed at high enough levels for clear separation of the cells that do not express this protein from normal cells. The GPI-anchored protein used in the selection must have been previously characterized, and methods to detect it in living cells be established. Because detection in most cases involves an antibody, the availability of this reagent will be critical to the mutant selection. Finally, depending on the selection strategy that is going to be used, it may be better to have a cell line that grows attached to a culture dish or in suspension culture.

In considering characterization of the mutant after it is isolated, an important factor in deciding what cell line to use is whether GPI biosynthesis can be measured in this cell. Elucidation of the specific step in the pathway that is affected in the mutant will require biochemical analysis of the various reactions. Therefore, whether it will be possible to assay each step in GPI biosynthesis in the chosen cell line should be established before beginning the mutant selection.

The choice of cell line will also vary, depending on what the mutant is to be used for. For instance, if a mutant in a specific reaction in GPI biosynthesis is sought, to be able to express clone cDNAs that encode proteins required for this step, then the cell line should be readily transfectable. Alternatively, if the mutant is isolated to probe the effect of the loss of GPI or a GPI-anchored protein in specific cellular events, this process must be measurable in the cell line chosen for the selection.

Finally, some special considerations should be made if a mutant in a particular step in GPI biosynthesis is sought. The PIG-A gene, which encodes a protein required for the first step in the pathway that was defined by the class-A mutant, is located on the X chromosome. Therefore, in diploid cells, disruption of any other gene required for GPI biosynthesis not on the X chromosome would require mutations of two copies, and would occur at a much higher frequency than class-A defects. Indeed, the selection of Thy-1-deficient mutants from murine lymphoma cells confirmed that class A defects were found at a much higher frequency than any other mutation ***(17)***. If other than a class A mutant is sought, one way around this problem is to use a cell line such as Chinese hamster ovary (CHO), which is functionally hemizygous. Because CHO cells express most proteins from only one allele ***(18)***, the probability of disrupting any gene in these cells should be about equal. However, mutant selections using CHO cells indicate that this argument is only partially true. Two separate selections for CHO cells defective in GPI biosynthesis have yielded no class A mutants ***(9,19)***, but the only mutant isolated in each case is defective in the second step of the pathway (class L). This finding suggests that the PIG-L gene is especially prone to disruption in these cells, and that CHO cells might not be useful for the isolation of mutants with defects in other reactions in GPI biosynthesis. Although a different cell line could be used to isolate novel mutants, an alternative solution to this problem would be to use CHO cells that have been transfected with the PIG-L cDNA, and to stably incorporate numerous copies of this sequence into their genome. Such a strategy has been suggested ***(9)***, but no evidence has yet been presented to indicate whether it will actually work.

1.2.2. Choice of a Mutagen

Most, but not all, of the mammalian cell mutants in GPI biosynthesis isolated to date have been generated with chemical mutagens. In general, mutagenesis before selection should increase the number of defective cells in the population and, therefore, the probability of isolating a mutant. The most common mutagen used to generate cells defective in GPI biosynthesis is ethylmethane sulfonate (EMS). This monofunctional ethylating agent is thought to generate point mutations ***(20)***. One mutant selection done with K562 cells used *N*-methyl-*N'*-nitro-*N*-nitrosoguanidine (MNNG), in an attempt to generate different mutations than previously found in EMS-treated cells ***(21)***. Although novel mutants were isolated, it is not clear whether the difference in mutagen is responsible for this. Therefore, whether the choice of mutagen really defines the defects that will be found can only be determined by further study.

1.2.3. Choice of a Selection Strategy

Using antibody detection of the GPI-anchored protein, the possible methods for selection of cells that have lost this protein from their surface are panning, complement-mediated lysis, and cell sorting by flow cytometry (FACS). Panning may be effective for isolating cells that have gained a surface protein, but it does not allow sufficient enrichment of the negative cells to make it useful for the isolation of mutants in GPI biosynthesis. Both complement-mediated lysis and FACS have been successfully used to isolate mutants. The choice between these methods may depend on the availability of reagents and equipment. FACS sorting is highly effective, but requires a flow cytometer, and works best with monoclonal antibodies. Complement-mediated lysis, on the other hand, requires no special equipment, and is more effective with polyclonal antibodies.

2. Materials

2.1. Electroporation

1. cDNA(s) for GPI-anchored protein(s) in a mammalian expression vector.
2. Selective agent (geneticin [G418, Gibco], hygromycin B [Sigma, St. Louis, MO], or Zeocin [Invitrogen], depending on resistance in vector).
3. Phosphate-buffered saline (PBS).
4. Growth medium.
5. Electroporator.
6. 2-mm gap electroporation cuvet.

2.2. Mutagenesis

1. Ethylmethane sulfonate (EMS, also called methanesulfonic acid ethyl ester [Sigma]).
2. Growth medium.

2.3. Complement-Mediated Lysis

1. Polyclonal antibody to GPI-anchored marker protein.
2. Complement serum (human serum from any source will work, in most cases).
3. Growth medium (serum-free).

2.4. Fluorescence-Activated Cell Sorting

1. Monoclonal antibody to GPI-anchored marker protein.
2. Fluorescein-conjugated secondary antibody.

3. Flow cytometer.
4. PBS.
5. Propidium iodide.

2.5. Cell Fusion

1. 50% Polyethylene glycol (PEG, Hybri-Max mol wt 3000–3700 [Sigma]) in L-glutamine-free DMEM. Prepare by microwaving the solution to boiling several times, until the PEG is dissolved.
2. PBS.
3. Glutamine-free culture medium.

3. Methods

In some cases, such as the transfection of genes for GPI-anchored proteins into the cell to serve as markers for selection of mutants, several techniques are involved, and the final result can be accomplished by numerous methods. One or two methods are presented for each procedure. However, omission of a technique does not mean that it cannot be used for the task described.

3.1. Preparation of Cells

If the cell line to be used naturally expresses one or more GPI-anchored protein that can be used to isolate cells that have lost this marker, then the only preparation necessary before beginning mutant selection is mutagenesis. However, if the GPI-anchored proteins must be introduced into the cell line chosen for mutant isolation, then stable transfectants expressing relatively high levels of the marker proteins must be prepared first.

3.1.1. Transfection of GPI-Anchored Protein(s) by Electroporation

1. Detach adherent cells from culture dish (use method typically used for that cell type) and harvest by centrifugation (5 min at 1000*g*) in 15-mL conical tubes (use centrifugation only with suspension cultures).
2. Wash the cells four times with PBS, pelleting them by centrifugation. Count the cells prior to the final centrifugation step, and resuspend them in PBS at a density of 1×10^7 cells/mL.
3. Cool the cells on ice for at least 30 min.
4. Transfer 400 µL of cells (4×10^6) into a prechilled 2-mm gap cuvet. Add 10 µg cDNA in minimal volume (<5 µL). Mix by gently flicking the cuvet.
5. Electroporate with appropriate voltage (varies with cell type). Place cuvet on ice for 10 min before manipulating cells.
6. Remove cells from cuvet and place in culture in appropriate medium.
7. After 24 h, add selective agent (i.e., geneticin, hygromycin B, Zeocin) to the culture at a concentration known to kill the cells being used.
8. Allow cells to grow until all the nontransfected cells have died, and colonies of cells are large enough to locate without a microscope (usually 2 to 3 wk). Change

medium as needed, maintaining the selective agent (i.e., geneticin, hygromycin B, Zeocin) at all times.

9. Clone single cells by limiting dilution. Characterize expression of GPI-anchored protein by FACS or enzymatic assay (if possible, e.g., can be used with alkaline phosphatase).

3.1.2. Mutagenesis

1. Grow cells to roughly 50% confluence. Remove culture medium and replace with fresh medium containing 100–400 µg/mL ethylmethane sulfonate (EMS).
2. After 24 h, remove the EMS-containing medium, and replace with normal culture medium.
3. Allow cells to recover for 24–48 h before beginning mutant selection.

3.2. Selection of Mutants

After mutagenesis, mutants defective in GPI biosynthesis are isolated by selecting cells that no longer express the GPI-anchored marker(s). Two methods for this, complement-mediated lysis and FACS, are described.

3.2.1. Complement-Mediated Lysis

1. Incubate EMS-mutagenized cells (>5 × 10^7/selection) with 1 mg/mL polyclonal antibody against the GPI-anchored marker protein in serum-free medium for 1 h at 37°C.
2. Remove the medium and replace it with medium containing 20% human serum. Incubate the cells at 37°C for 1 h, during which the complement components in the serum assembled to form the membrane attack complex which kill the GPI-anchored marker-expressing cells.
3. Allow the surviving cells to repopulate the flask, and then subject them to 2–3 more rounds of complement-mediated lysis (until few cells are killed by the antibody combination).
4. Clone the surviving cells by limiting dilution.
5. Characterize expression of GPI-anchored protein by FACS or enzymatic assay, to determine which clones no longer express the GPI-anchored marker on their surface.

3.2.2. Fluorescence-Activated Cell Sorting

1. Detach adherent cells from culture dish and harvest by centrifugation (5 min at 1000*g*; use centrifugation only with suspension cultures). Proteolytic methods (e.g., trypsin) can be used to detach the cells, but should be avoided if a nonprolytic method is available.
2. Aliquot 1 × 10^6 cells/tube into 12 × 75 plastic tubes.
3. Incubate cells for 45 min at 4°C in PBS containing 1% serum and monoclonal antibody against the GPI-anchored marker protein (the concentration should be determined experimentally).

4. Fill the tubes with PBS plus 1% serum and pellet cells by centrifugation (1000*g*, 5 min). Resuspend the cells in PBS, plus 1% serum containing fluorescein-conjugated secondary antibody (concentration should be determined experimentally by optimizing signal). Incubate at 4°C for 45 min.
5. Pellet cells by centrifugation (1000*g*, 5 min), and resuspend in 0.5 mL PBS containing 1 µg/mL propidium iodide.
6. Analyze by flow cytometry. Gate out the dead (propidium iodide-positive) cells, and collect the cells (appropriately 1% of the total) with the lowest fluorescence from the GPI-anchored protein antibody.
7. Allow the selected cells to grow for several days, and repeat the cell sorting two to three times.
8. Clone the cells selected in the final round by limiting dilution.
9. Characterize expression of GPI-anchored protein by FACS or enzymatic assay, to determine which clones no longer express the GPI-anchored marker on their surface.

3.3. Identification of Affected Step in GPI Biosynthesis

Newly selected mutants in GPI-anchored proteins can be characterized in two ways. Complementation analysis can be done to determine if the mutant has a defect that is the same as the previously characterized mutants in this pathway. Because the affected step is known for each complementation group described to date, the finding that the isolated mutant belongs to one of these classes will reveal the reaction that is defective. A new mutant can also be characterized biochemically, to determine which biosynthetic reaction is affected.

3.3.1. Complementation Analysis

1. Detach cells from culture dish and harvest by centrifugation (1000*g*, 5 min). Wash the cells one time with PBS by centrifugation.
2. Combine 2.5×10^6 of each type to be fused in L-glutamine-free medium.
3. Pellet the cells by centrifugation, and resuspended in 1 mL of 50% polyethylene glycol (PEG) in L-glutamine-free medium over a 1-min time period, with constant mixing.
4. Dilute this solution by the addition of 1 mL of L-glutamine-free medium over 1 min, followed by 5 mL of medium containing 10% serum over 3 min.
5. Centrifuge (1000*g*, 5 min) the cell solution. Resuspend the cells in 5 mL of medium containing 10% serum in a 100-cm^2 culture dish, and incubated at 37°C for at least 24 h.
6. Assess complementation by determining the percentage of fused cells that express the GPI-anchored marker(s).

3.3.2. Analysis of GPI Biosynthesis

The first three steps in GPI biosynthesis in mammalian cells can be measured in vitro, using a cell lysate or microsomal preparation. The details of these methods are presented in Chapter 3 of this volume. The absence of any of the first three intermediates made in an incubation with UDP-[^{3}H]GlcNAc will indicate the reaction that is affected in the mutant.

Defects in the later steps in GPI biosynthesis should be detected by in vitro labeling of these GPI intermediates with [^{3}H]mannose, as described by Ueda et al. ***(22)*** or Kamitani et al. ***(23)***. As in the in vitro analysis, the absence of one or more mannose-containing intermediates will indicate which step in the pathway is defective in the mutant.

4. Notes

1. Because the selection method targets cells that no longer express a GPI-anchored protein on their surface, mutants in which the gene for this marker protein has been affected will be selected, along with the desired mutants in GPI biosynthesis. These mutants can be avoided by using two GPI-anchored marker proteins. By requiring the surface expression of both proteins be affected, cells with defects in one of the markers will not be selected.
2. Both FACS and complement-mediated lysis can be done with adherent and suspension cultures. However, it is usually more convenient to use cells grown in suspension with FACS and adherent cells for complement-mediated lysis. The isolation of colonies derived from a single cell is also more straightforward with adherent cells.
3. The concentration of EMS used should balance the mutagenic and cytotoxic effects of this compound on cells. The optimal concentration can be determined with the cell line being used, by measuring these parameters with different doses of EMS. This optimum often occurs at the concentration of EMS that results in 50% cytotoxicity. Therefore, this measure can be used to determine the appropriate EMS concentration, if characterization of the mutation frequency is not done.
4. The surface expression of the GPI-anchored marker protein should be assessed after transfection and selection of mutants. If FACS is available, and is being used in the selection, it is probably easiest to use this to quantify expression. However, alternative methods, such as biochemical, histochemical, and immunochemical assays, will work equally well for this purpose.
5. In the complementation analysis, it is not necessary to eliminate the unfused cells from the analysis, because they will not show any expression of the GPI-anchored marker protein. Fused cells can be identified by their large size, which, if there is complementation of the mutation, should be positive a high percentage of the time.

References

1. Chapman, A., Fujimoto, K., and Kornfeld, S. (1980) The primary defect in class e Thy-1- negative mutant mouse lymphoma cells is the inability to synthesize dolichol-P-mannose. *J. Biol. Chem.* **255,** 4441–4446.

2. Menon, A. K., Mayor, S., and Schwarz, R. T. (1990) Biosynthesis of glycosylphosphatidyl-inositol lipids in *Trypanosoma brucei:* involvement of mannosylphosphoryldolichol as the mannose donor. *EMBO J.* **9,** 4249–4258.
3. Urakaze, M., Kamitani, T., DeGasperi, R., Sugiyama, E., Chang, H., and Yeh, E. T. H. (1992) Identification of a missing link in glycosylphosphatidylinositol anchor biosynthesis in mammalian cells. *J. Biol. Chem.* **267,** 6459–6462.
4. Inoue, N., Kinoshita, T., Orii, T., and Takeda, J. (1993) Cloning of a human gene, PIG-F, a component of glycosylphosphatidylinositol anchor biosynthesis, by a novel expression cloning strategy. *J. Biol. Chem.* **268,** 6882–6885.
5. Miyata, T., Takeda, J., Iida, Y., Yamada, N., Inoue, N., Takahashi, M., et al. (1993) The cloning of PIG-A, a component in the early step of GPI- anchor biosynthesis. *Science* **259,** 1318–1320.
6. Kamitani, T., Chang, H.-M., Rollins, C., Waneck, G. L., and Yeh, E. T. H. (1993) Correction of the class H defect in glycosylphosphatidylinositol anchor biosynthesis in Ltk$^-$ cells by a human cDNA clone. *J. Biol. Chem.* **268,** 20,733–20,736.
7. Inoue, N., Watanabe, R., Takeda, J., and Kinoshita, T. (1996) PIG-C, one of the three human genes involved in the first step of glycosylphosphatidylinositol biosynthesis is a homologue of Saccharomyces cerevisiae GPI2. *Biochem. Biophys. Res. Commun.* **226,** 193–199.
8. Kinoshita, T., Takahashi, M., Inoue, N., Miyata, T., and Takeda, J. (1994) Expression cloning of genes for GPI-anchor biosynthesis. *Brazilian J. Med. Biol. Res.* **27,** 127–132.
9. Nakamura, N., Inoue, N., Watanabe, R., Takahashi, M., Takeda, J., Stevens, V. L., and Kinoshita, T. (1997) Expression cloning of PIG-L, a candidate N-acetylglucosaminylphos-phatidylinositol deacetylase. *J. Biol. Chem.* **272,** 15,834–15,840.
10. Lazar, D. F., Knez, J. J., Medof, M. E., Cuatrecasas, P., and Saltiel, A. R. (1994) Stimulation of glycogen synthesis by insulin in human erythroleukemia cells requires the synthesis of glycosyl-phosphatidylinositol. *Proc. Natl. Acad. Sci. USA* **1,** 9665–9669.
11. Leidich, S. D., Drapp, D. A., and Orlean, P. (1994) A conditionally lethal yeast mutant blocked at the first step in glycosyl phosphatidylinositol anchor synthesis. *J. Biol. Chem.* **269,** 10,193–10,196.
12. Kawagoe, K., Kitamura, D., Okabe, M., Taniuchi, I., Ikawa, M., Watanabe, T., Kinoshita, T., and Takeda, J. (1996) Glycosylphosphatidylinositol-anchor-deficient mice: implications for clonal dominance of mutant cells in paroxysmal nocturnal hemoglobinuria. *Blood* **87,** 3600–3606.
13. Trowbridge, I. S., Hyman, R., and Mazauskas, C. (1978) The synthesis and properties of T25 glycoprotein in Thy-1-negative mutant lymphoma cells. *Cell* **14,** 21–32.
14. Conzelmann, A., Spiazzi, A., Bron, C., and Hyman, R. (1988) No glycolipid anchors are added to Thy-1 glycoprotein in Thy-1-negative mutant thymoma cells of four different complementation classes. *Mol. Cell. Biol.* **8,** 674–678.

15. Fatemi, S. H. and Tartakoff, A. M. (1988) The phenotype of five classes of T lymphoma mutants. Defective glycophospholipid anchoring, rapid degradation, and secretion of Thy-1 glycoprotein. *J. Biol. Chem.* **263,** 1288–1294.
16. Fatemi, S. H. and Tartakoff, A. M. (1986) Hydrophobic anchor-deficient Thy-1 is secreted by a class e mutant T lymphoma. *Cell* **46,** 653–657.
17. Hyman, R. (1988) Somatic genetic analysis of the expression of cell surface molecules. *Trends Gen.* **4,** 5–8.
18. Worton, R. G. (1978) Karyotypic heterogeneity in CHO cell lines. *Cytogenet. Cell Genet.* **21,** 105–110.
19. Stevens, V. L., Zhang, H., and Harreman, M. (1996) Isolation and characterization of Chinese hamster ovary (CHO) mutant defective in the second step in glycosylphosphatidylinositol biosynthesis. *Biochem. J.* **313,** 253–258.
20. Sega, G. A. (1984) A review of the genetic effects of ethyl methanesulfonate. *Mutat. Res.* **134,** 113–142.
21. Mohney, R. P., Knez, J. J., Ravi, L., Sevlever, D., Rosenberry, T. L., Hirose, S., and Medof, M. E. (1994) Glycoinositol phospholipid anchor-defective K562 mutants with biochemical lesions distinct from those in Thy-1$^-$ murine lymphoma mutants. *J. Biol. Chem.* **269,** 6536–6542.
22. Ueda, E., Sevlever, D., Prince, G. M., Rosenberry, T. L., Hirose, S., and Medof, M. E. (1993) A candidate mammalian glycoinositol phospholipid precursor containing three phosphoethanoamines. *J. Biol. Chem.* **268,** 9998–10,002.
23. Kamitani, T., Menon, A. K., Hallaq, Y., Warren, C. D., and Yeh, E. T. H. (1992) Complexity of ethanolamine phosphate addition in the biosynthesis of glycosylphosphatidylinositol anchors in mammalian cells. *J. Biol. Chem.* **267,** 24,611–24,619.

3

Analysis of the Cell-Surface Distribution of GPI-Anchored Proteins

Satyajit Mayor

1. Introduction

1.1. Background

Many cell-surface eukaryotic proteins, including several receptors, enzymes, and adhesion molecules, have a glycolipid modification at their carboxy-terminal end **(Fig. 1A)**. This is a posttranslational modification that serves as a membrane anchor and involves the replacement of the carboxy-terminal peptide sequence of the protein by a glycosylphosphatidylinositol (GPI) moiety ***(1,2)***. The structure and biosynthesis of the GPI moiety are now well understood ***(1–4)***. In spite of the extensive biochemical information on the GPI moiety of GPI-anchored proteins, the functions of this ubiquitous protein modification are less understood, although it has been implicated in a variety of cell biological processes ***(5)***. The GPI anchor has been proposed to act as an apical targeting signal for proteins in some epithelial cell types via its association with putative glycolipid rafts in the *trans*-Golgi network ***(6,7)***. GPI-anchoring has also been shown to be important for the intracellular signaling capacity of several proteins, especially in lymphocytes. In most cases, the crosslinking of the protein is a prerequisite for their signaling function ***(8,9)***.

In immunolocalization studies, GPI-anchored proteins have been found to be clustered at the cell surface, and a significant fraction of the clusters are localized to 50–60 nm caveolin/VIP-21-coated membrane invaginations called caveolae ***(10–12)***. In these and other studies, caveolae localization of GPI-anchored proteins was determined by immunolocalization of these proteins using primary (polyclonal or monoclonal) antibodies followed by labeled polyclonal secondary IgGs. Considering that GPI-bearing proteins are lipid-anchored, it is possible that multiple layers of antibodies may cause a redistri-

From: *Methods in Molecular Biology, Vol. 116: Protein Lipidation Protocols*
Edited by: M. H. Gelb © Humana Press Inc., Totowa, NJ

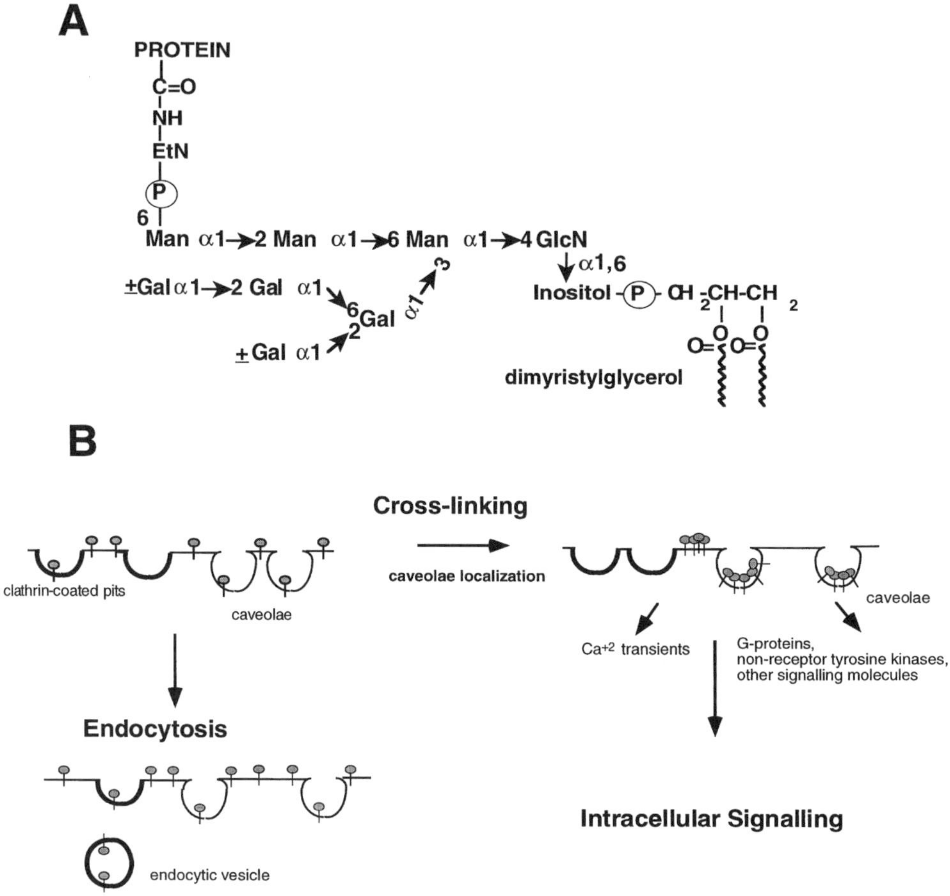

Fig. 1. **(A)** Complete structure of a GPI-anchor present on a variant-surface glycoprotein from *Trypanosoma brucei*. Structure drawn according to **ref.** ***(13)***. **(B)** Dynamics of GPI-anchored proteins at the cell surface. A schematic of the cell-surface dynamics of GPI-anchored proteins wherein in their native-state GPI-anchored proteins are free to diffuse in the plane of the bilayer. Consequently, they are not enriched nor depleted in coated or noncoated pits, and are taken into the cell by mechanisms of bulk membrane endocytosis. On crosslinking (by antibodies or by physiological agents), these proteins will be clustered and preferentially localize to caveolae.

bution of the molecules in the plane of the membrane. Furthermore, the extent of fixation of these proteins in the plane of the membrane may differ from that of transmembrane proteins with proteinaceous tails. In general, it has been observed that the fixation of lipids in membranes by routine fixation methods (especially formaldehyde fixation) has been very unreliable *(14)*. To avoid such potential artifacts, Mayer et al. *(15)* directly conjugated fluorophores to monoclonal antibodies or Fab molecules, and observed the "native" distribution of GPI-anchored proteins in living and fixed cells. Using such methods, Mayer et al. *(15)* have investigated the ability of these proteins to remain mobile in the plane of the membrane after routine fixation conditions. They found that GPI-anchored proteins are diffusely distributed at the cell surface and, even after fixation, are prone to redistribution after crosslinking with secondary antibodies *(15)*. These data led to the conclusion that GPI-anchored proteins are not constitutively concentrated in caveolae; they are enriched in these structures only after crosslinking with polyclonal secondary antibodies *(15)*. Similar results have been observed by the use of fluorescein-modified folic acid as a monovalent ligand for the GPI-anchored folate receptor, showing that the diffuse distribution of these proteins is not a consequence of antibody binding *(16)*. Analyses of the cell-surface distribution of GPI-anchored folate receptor and alkaline phosphatase by electron microscopy have also confirmed that these proteins are not constitutively enriched in caveolae *(15,17)*. Thus, it appears that multimerization of GPI-anchored proteins regulates their sequestration in caveolae, but in the absence of agents that promote clustering, they are diffusely distributed over the plasma membrane *(15)* (**Figs. 1B** and **2**). However, physiological clustering agents for GPI-anchored proteins have yet to be identified.

These observations introduce a cautionary note in the age-old adage "seeing is believing," and it follows that "how we see" may also informs us about the system under observation. In this chapter, methods that will permit the observation of "native" distribution of GPI-anchored proteins at the cell surface by fluorescence and electron microscopic techniques will be outlined.

1.2. Analysis of the Distribution of GPI-Anchored Proteins by Fluorescence Microscopy

Fluorescence microscopy may be used to observe the native distribution of GPI-anchored proteins at the cell surface at the light (fluorescence) microscopy level, provided adequate precautions are taken to prevent the redistribution of these molecules during the observation procedure. Preferably, monovalent reagents for the GPI-linked proteins must be used on living cells or fixed tissue sections. The fixation conditions will depend on the nature of tissue or cell preparations, and the GPI-anchored protein in question. In the absence of

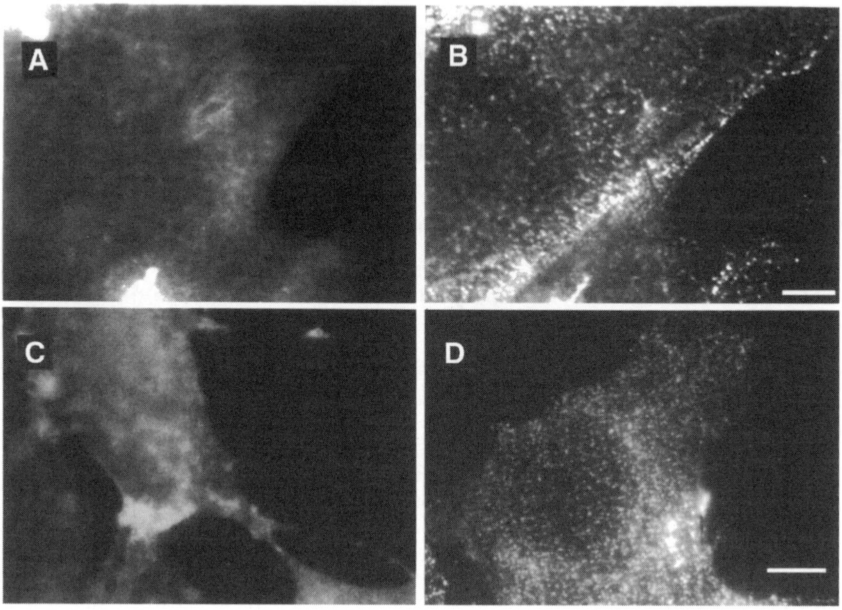

Fig. 2. Antibody-induced redistribution of GPI-anchored proteins in living and fixed cells. **(A, B)** 3T3-L1 cells transfected with folate receptor (G3G2 cells) were labeled with Cy3-conjugated MAb (10 μg/mL) to folate receptor (Cy3-MOv19). The cells were then imaged before (A) and 12 min after (B) the addition of unlabeled polyclonal goat IgG against mouse mAb (final concentration 20 μg/mL) to the cells as described. Mouse MAb to the human folate receptor (MOv19 ***[18]***) a gift from Centocor Corp., Malvern, PA. **(C, D)** 3T3-L1 cells were fixed for 20 min at room temperature with 3% paraformaldehyde prior to staining with FITC-anti-Thy-1 (4 μg/mL; Gibco-BRL) (C). The diffuse distributions of FITC-anti-Thy could be converted to a clustered distribution (D) even if the cells had been fixed for 20 min at 23°C in 3% paraformaldehyde prior to addition of a crosslinking polyclonal secondary IgG (20 μg/mL for 20 min. at room temperature). This shows that GPI-anchored proteins are able to redistribute in paraformaldehyde-fixed cells. The ability of GPI-anchored proteins to redistribute after fixation presumably accounts for the previous immunolocalizations of these proteins in tight clusters ***(11,12)***. This redistribution was not observed in cells that had been fixed for longer periods at room temperature (>1 h), or fixed in the presence of 0.1–0.2% glutaraldehyde along with 3% paraformaldehyde. Bar = 10 μm.

monovalent reagents, adequate fixation conditions must be worked out that will allow the observation of the native distribution of these proteins without altering their antigenicity and their "native distribution." A method that has been successfully applied to study the native distribution of GPI-anchored proteins on many different cell types will be described. However, these methods must be taken as starting points for the study of other GPI-anchored proteins in different contexts.

1.3. Quantitative Analysis of the Distribution of GPI-Anchored Proteins at the Cell Surface by Electron Microscopy

Fluorescence microscopy provides qualitative analyses of the spatial distribution of GPI-anchored proteins at a resolution of half the wavelength of light (~250–300 nm). However, if GPI-anchored proteins are distributed in clusters at an intercluster distance <250 nm, then light microscopy will not be able to detect any heterogeneity in the distribution of the fluorophores if the clusters are uniformLy distributed. On the other hand, immunoelectron microscopy (e.g., immunogold labeling techniques) greatly enhances intercluster resolution by using small sized (1–40 nm) electron-dense particles, very well resolved by the electron beam. A potential source of error in these analyses is owing to the inherent insensitivity of the technique. Labeling efficiencies of >1% are rarely achieved because of steric and other considerations, thereby making the quantitation of the number of molecules per labeling event (gold particle detected) quite difficult ***(14)***.

2. Materials

2.1. Materials Required for the Qualitative Analysis of the Distribution of GPI-Anchored Proteins by Fluorescence Microscopy

1. MAb: Monoclonal antibodies or monovalent ligands are usually the reagents of choice. These must be obtained for specific GPI-anchored proteins in question. To ensure monovalent binding to the specific protein (in the absence of definitive information of the valency of interaction), Fab fragments of MAbs may be generated by papain digestion followed by protein A removal of intact and cleaved Fc portions according to published procedures ***(19)***. Fab fragments of MAbs must be checked for contamination with intact antibodies by SDS gel electrophoresis and Coomassie staining, and shown to be free of contaminating intact antibody even on heavily overloaded gels.
2. Poly-D-lysine-treated cover slip bottom dishes are made according to previously published procedures ***(20)*** or obtained from Costar, Inc. and used for growing cells for all microscopy studies. Alternatively, cells may be directly grown on cover slips placed in the bottom of a 35-mm dish.
3. Glutaraldehyde (EM grade, 8% aqueous solution) is from EM Sciences (Fort Washington, PA) and is stored in sealed vials at 4°C.
4. Paraformaldehyde is prepared as a 30% solution (w/v) in water. Thirty grams of paraformaldehyde (Sigma Chemical Co., St. Louis, MO) are dissolved in water at 60°C, and a few drops of 1 *N* NaOH added until the powder dissolves. The final solution is made up to a volume of 100 mL. The resultant solution must be

stored at –20°C. Working concentrations are always prepared by thawing a fresh aliquot at 60°C and making up the desired strength in the appropriate buffer.

5. Medium 1 (150 m*M* NaCl, 20 m*M* HEPES, pH 7.4, 5 m*M* KCl, 1 m*M* $CaCl_2$, 1 m*M* $MgCl_2$) is prepared as a 5X stock and stored at room temperature.
6. Bovine serum albumin (BSA, RIA Grade, Sigma Chemical Co.) solutions (1% and 5%) are freshly prepared in medium 1.
7. Fluorophores: Cy3 (Amersham Life Science, Buckinghamshire, UK), fluoroscein succinimidyl ester, and Texas red succinimidyl ester, may be obtained from Molecular Probes.
8. All other chemicals are from Sigma Chemical Co. (St. Louis, MO) and tissue culture supplies were from GIBCO-BRL (Gaithersburg, MD) unless otherwise specified.

2.2. Materials Required for the Quantitative Analysis of the Distribution of GPI-Anchored Proteins by Electron Microscopy

1. $K_3Fe(CN)_6$, osmium tetroxide, uranyl acetate, lead citrate, and EM-bed 812 resin are purchased from EM Sciences. Aqueous hydrofluoric acid (50%) is purchased from E. Merck (Germany).
2. 0.2 *M* sodium cacodylate (EM Sciences), pH 7.4, buffer is used as a 2X stock to make up different postfixation solutions.
3. Glutaraldehyde and tannic acid fixative: 2% solution of tannic acid (EM Sciences) is prepared by dissolving 20 mg in 1 mL 0.2*M* sodium cacodylate and adjusted to pH 7.4. The resulting solution is filtered through a 0.2-μ*M* filter before adding glutaraldehyde (from an 8% solution in water) to make the final fixative 1% glutaraldehyde and 1% tannic acid in 0. 1 *M* sodium cacodylate, pH 7.4. $CaCl_2$ is added for a final concentration of 3.4 m*M* $CaCl_2$. The fixative is prepared just prior to use, since tannic acid will begin to precipitate on storage.
4. Species-specific polyclonal IgGs to primary antibodies are obtained from Pierce Chemical Co. (Rockford, IL) or ICN Biomedicals, Inc. (Costa Mesa, CA). To remove any aggregated IgGs, the solution of antibodies may be pelleted by spinning the antibody suspension at 100,000*g* for 30 min in a table top ultracentrifuge. Usually this is not a problem for freshly reconstituted antibody solutions.
5. Gold-conjugated antibodies were obtained from Amersham Life Science. This manufacturer provides cluster analyses of the gold particles coated with antibodies.
6. 10X ocular eyepiece (Zeiss, Germany).

3. Methods (*see* Note 1)

3.1. Fluorescent Labeling of MAbs

Mabs or Fabs may be labeled with fluorophores such as Cy3 (Biological Detection Systems, Inc., Pittsburgh, PA), fluorescein, or Texas red (Molecular Probes, Eugene, OR) via succinimidyl esters, according to the manufacturer's instructions. Care should be taken to ensure that the labeling of reagent does not alter the binding characteristics for the protein in question. For antibody labeling, dye to protein ratios of 3 or 4 are usually optimal.

3.2. Cell Culture

Cells are grown in 25-cm^2 flasks in appropriate media and transferred to poly-D-lysine coated cover slip bottom dishes (*see* **Subheading 2.1.2.**) 3 d prior to the labeling of cell surface proteins. Cells are taken for labeling when they are about 60–70% confluent on the coverslip.

3.3. Labeling of Cells

Live or fixed cells are labeled with saturating concentration of reagents in medium 1 with 1% BSA for 1 h at 0°C or at room temperature, respectively. The level of saturation binding of the reagents used must be previously determined by labeling cells with increasing concentration of reagents and taking the cells through steps in **Subheadings 3.4.** and **3.5.**, and selecting the lowest concentration of reagents that generates the strongest fluorescence signal. The cells are then rinsed five times in medium 1 with 1% BSA to remove unbound reagents.

3.4. Fixation of Cells

Cells rinsed in medium 1 are incubated with 3% paraformaldehyde and glutaraldehyde (0.1–0.5%) for 30 min at room temperature. The exact concentration of the glutaraldehyde fixative may depend on the extent to which the antigenicity of the specific protein or the binding capacity of the specific protein is altered. If glutaraldehyde is left out, considerable redistribution of the GPI-anchored proteins is observed during treatment with crosslinking secondary antibodies **(Fig. 2)**. The cells are then rinsed in medium 1 and excess aldehyde groups of the fixative are quenched by incubating the cells with 25 m*M* NH_4Cl in medium 1 for 20 min at room temperature, followed by a 5-min incubation in medium 1 with 5% BSA.

3.5. Fluorescence Microscopy

Cells (live or fixed) are taken for fluorescence microscopy and observed with an oil immersion 60X objective on an inverted microscope equipped with a CCD camera to record images using the correct optical filters for different fluorophores. We have used a Leitz Diavert fluorescence microscope equipped with a Photometrics (Tucson, AZ) cooled CCD camera and driven by software from Inovison Corp. (Durham, NC) on a SPARC station 4/330 computer system (Sun Microsystems Inc., Mountainview, CA) or a Nikon TMD fluorescence microscope equipped with a Princeton Instruments (Princeton, NJ) cooled CCD camera driven by Metamorph software (Universal Imaging Corporation, Westchester, PA). For output purposes, the digital images are transferred to a Macintosh Power PC and printed via Adobe Photoshop software on a dye-sublimation printer (Phaser IIsdx, Tektronix Inc., Wilsonville,

OR). Images may also be recorded on film. Generally 400–1000 ASA is preferable for maximum sensitivity. Examples of fluorescence images obtained by these methods are shown in **Fig. 2**.

3.6. Cell Culture and Labeling for Electron Microscopy

Cells are grown on cover slip bottom dishes and labeled (usually live) with MAbs to respective GPI-anchored proteins as described in **Subheading 3.1.** (*see* **Note 2**).

3.7. Fixation Procedures for Electron Microscopy

The cells are then fixed with 3% paraformaldehyde and glutaraldehyde (0.1%–0.5%) for 30 min at room temperature (*see* **Subheading 3.4.**) and incubated in medium 1 containing 25 m*M* NH_4Cl for 20 min, and then blocked in medium 1 containing 10% FBS for 20 min. All incubations on fixed cells are carried out at room temperature.

3.8. Immunogold Labeling

The blocked preparations are then sequentially incubated in polyclonal rabbit anti-mouse antiserum (20 μg/mL specific immunoglobulin concentration; ICN Biomedicals, Inc., Costa Mesa, CA) followed by 1/30 dilution of 10 nm gold-conjugated antibodies. The cells are rinsed in medium 1 containing 10% FBS (3 × 5 min) between incubations, then in medium 1 without FBS (3 × 1 min), and transferred to 0.1 *M* sodium cacodylate, pH 7.4.

3.9. Postfixation Procedures

The cells are postfixed with glutaraldehyde and tannic acid fixative for 30 min, and then rinsed in 0.1 *M* sodium cacodylate (3 × 1 min).

3.10. Osmication and Epon Embedding

The cells are then osmicated (1 h at room temperature) in freshly prepared osmium-ferricyanide mixture (1% O_sO_4, 1.5% $K_3Fe(CN)_6$, in 0.1 *M* sodium cacodylate), rinsed (3 × 5 min) in cacodylate buffer and finally water. The preparations are then dehydrated in graded ethanol–water washes (5, 30, 50, 90, and 100% ethanol concentration) for a duration of 3 min in each ethanol concentrations. Then they are embedded in epon (EM-bed 812; EMS) first as a 1:1 mixture with ethanol (1 h at room temperature) and then as a complete mixture (1 h at room temperature). Adequate quantities (~1 mL/cover slip surface) of the epon polymer (including accelerant) must be made so that some of it may be used to make the cutting blocks as described below **(Fig. 3)**. Store the premixed epon at 4°C until further use.

Detailed procedures for the use of Epon resins (EM-bed 812 adequately substitutes for the original Shell polymer, Epon 812) may be obtained in **ref.** ***(21)***.

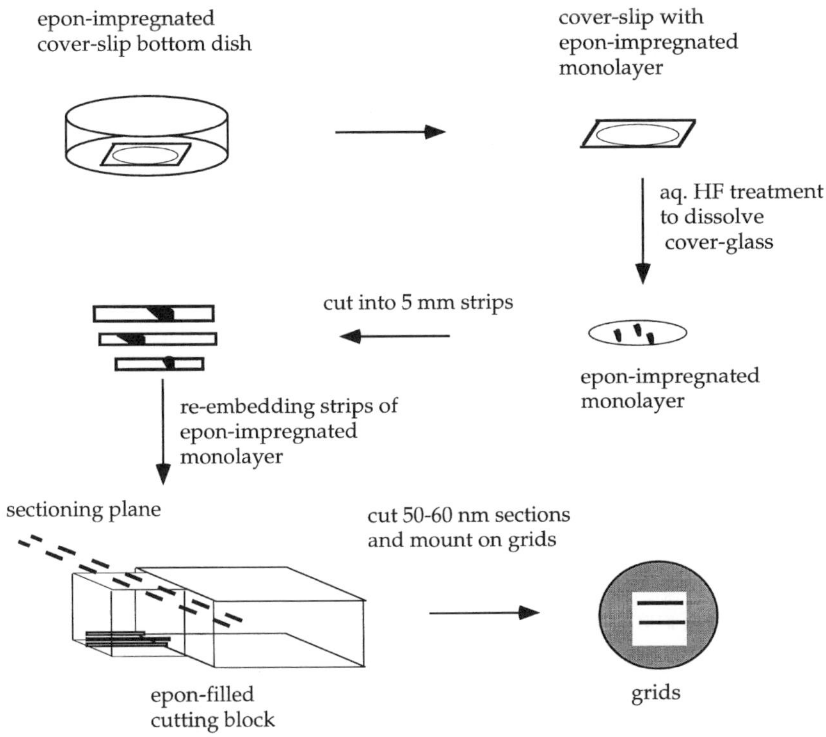

Fig. 3. Schematic for the processing of samples to quantitate surface distributions of gold particles by electron microscopy.

3.11. Cover Glass Removal

The glass cover slip along with the epon-soaked monolayer is carefully removed from each cover slip bottom dish and baked in an oven (24 h, 60°C). The cover glass is then dissolved in aqueous hydrofluoric acid (HF). The aqueous HF-treated plastic is washed extensively until there is no longer an acrid odor of the acid, and then the epon-embedded monolayers are dried in an oven at 60°C to remove all moisture thoroughly.

3.12. Preparation of Cutting Blocks

The dried epon-embedded plastic containing the monolayers is cut lengthwise into thin 5-mm-wide strips. Three or four of these strips may be re-embedded in cutting blocks using the same epon as above. The flat surface of the strips must be arranged parallel to the bottom of the block base. Care must be taken to prevent air bubbles between strips. If the strips begin to bend or buckle, this is because of inadequate removal of moisture from the plastic at the time of drying after HF treatment and subsequent washing.

3.13. Sectioning

Fifty- to 60-nm thick sections are cut from the these blocks on an RMC MT-7000 ultramicrotome perpendicular to the plane of the cell monolayer. The sections are then poststained in 4% uranyl acetate for 15 min followed by 0.4% lead citrate for 4 min prior to observation on a JEOL JEM-1200 EXII electron microscope at 80 kV. All images for the statistical data are taken at a magnification of 20,000 ×, collected at random, and contact printed on 7.5 × 10 cm negatives.

3.14. Quantitative Analyses

Quantitative analyses is carried out using a 10X ocular lens. More than 50 negatives of randomly selected fields are analyzed for each condition in each piece of experiment. Several criteria must be established so that the labeling with the gold particles may be quantified (*see* **Note 3**). Membrane length is estimated with a scale at a resolution of 0.1 cm (1 cm = 500 nm), and considered as the length of uninterrupted membrane where a definite bilayer may be clearly discerned. Membrane protrusions are also noted, but not included in the analyses of undifferentiated areas of the cell membrane. Areas where a definite bilayer is not observed are discarded and not considered as part of the analyses. Each gold particle or cluster on a *bona fide* membrane surface is counted as a "labeling event," and its properties are tabulated in terms of its association with a given type of plasma membrane. A cluster contains more than three gold particles separated by <50 nm from the nearest neighbour. The type of plasma membrane is categorized in terms of whether it is a membrane invagination with a discernible coat or not, undifferentiated membrane, or membrane protrusion. Gold particles (or clusters) <50 nm from the edge of caveolae (50-nm invaginations) or coated pits (coated membrane invaginations >70 nm) are considered to be associated with them. Quantitative parameters are then determined for each experimental condition as below:

1. Expected gold particle density over caveolae or coated pits = gold particle density/U length × total length of membrane associated with these structures.
2. Relative concentration = the ratio of the observed gold particle density in caveolae or coated pits to the expected gold particle density over these structures.

3.15. Nonspecific Binding of Gold Particles to Cells

This is determined by using an isotype-matched IgG as the primary antibody, and the labeling procedures and analytical methods as outlined above.

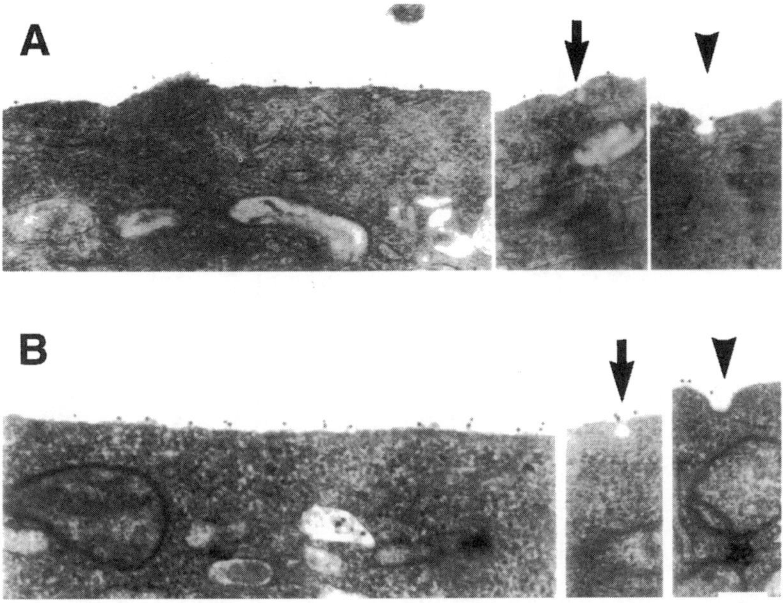

Fig. 4. Electron microscopic analyses of cell-surface distribution of GPI-anchored proteins. MA104 cells **(A)** expressing the folate receptor were incubated with MOv19 (2 µg/mL), and CHO cells expressing to decay accelerating factor (DAF) **(B)** were incubated mouse MAb against DAF, 1A10 (2 µg/mL), for 1 h at 4°C, rinsed, and incubated for an additional 10 min at 37°C. The cells were then fixed with 3% paraformaldehyde and 0.5% glutaraldehyde for 30 min at room temperature prior to labeling with rabbit polyclonal serum against mouse IgG (20 µg/mL) followed by gold-(10 nm) labeled goat antirabbit IgG (2 µg/mL) and processed for electron microscopy. Mouse mAb to DAF, 1A10 *(22)* was a gift of M. Davitz (New York University). Arrows and arrowheads indicate structures identified as caveolae and coated pits, respectively. Bar = 200 nm.

3.16. Fluorescence Images

These are shown in **Fig. 2** , whereas **Fig. 4** and **Table 1** represent a sample analyses by electron microscopy.

4. Notes

1. **Subheading 3** provides separate procedures for the analysis of the cell-surface distribution of GPI-anchored protein by fluorescence microscopy (**Subheadings 3.1.–3.5.**) and electron microscopy (**Subheadings 3.6.–3.15.**).
2. The immunolabeling procedures outlined here may be used on fixed tissue samples as well. However, the labeling efficiencies for the primary antibody are even lower that those obtained in these analyses.

Table 1
Quantative Analyses of the Distribution of GPI-Anchored Proteins at the Cell Surface[a]

		Number of				% Gold associated with	
Cells, GPI-anchored protein	Membrane length (μm)	Caveolae	Coated pits	Number of gold particles	% Gold clusters	Caveolae, relative conc.[b]	Coated pits relative conc.[b]
MA104 cells	228	37	40	488	3	4.7	5.4
folate receptor						(1.9)	(1.2)
CHO cells	174	41	28	926	9	3.6	5.6
(DAF)						(1.0)	(1.4)

[a]Cells were processed for electron microscopy as described in **Fig. 4**. Data in this table are adapted from Table 1 in **ref. *16***. *See* **Subheadings 3.6.–3.15.** for an explanation of the quantitative methodology.

[b]The average membrane length associated with caveolae and coated pits was 0.15 and 0.25 μm, respectively.

3. The detection efficiencies for fixed primary antibodies are not very different from those obtained for unfixed primary antibodies.

Acknowledgments

The author thanks F. R. Maxfield (Cornell University Medical College, New York) for his generous support, and Kristy Brown (Columbia University, New York) for her invaluable assistance with electron microscopy. The author is grateful to the International Human Frontiers Science Program for a research grant (RG-328/96), and the National Centre for Biological Sciences for financial support. The author would also like to thank F. F. Bosphorus and C. Head for inspiration.

References

1. Englund, P. T. (1993) The structure and biosynthesis of glycosyl phosphatidylinositol protein anchors [Review]. *Annu. Rev. Biochem.* **62,** 121–138.
2. Field, M. C. and Menon, A. K. (1992) Glycolipid-anchoring of cell surface proteins, in *Lipid Modification of Proteins* (Schlesinger, M. J., and Schlesinger, M. J. S., eds.), CRC, Boca Raton, FL, pp. 83–134.
3. Mayor, S. and Menon, A. K. (1990) Structural analysis of the glycosylinositol phospholipid anchors of membrane proteins. *Methods: A Companion to Methods in Enzymol.* **1,** 297–305.
4. McConville, M. J. and Ferguson, M. A. (1993) The structure, biosynthesis and function of glycosylated phosphatidylinositols in the parasitic protozoa and higher eukaryotes. *Biochem. J.* **294,** 30–204.
5. Ferguson, M. A. J. (1994) What can GPI do for you. *Parasitol. Today* **10,** 48–52.
6. Lisanti, M. P. and Rodriguez-Boulan, E. (1990) Glycophospholipid membrane anchoring provides clues to the mechanism of protein sorting in polarized epithelial cells. *Trends Biochem. Sci.* **15,** 113–118.
7. Simons, K. and Ikonen, E. (1997) Functional rafts in membranes. *Nature* **387,** 569–570.
8. Brown, D. (1993) The tyrosine kinase connection: how GPI-anchored proteins activate T cells. *Curr. Opinion Immunol.* **5,** 349–354.
9. Robinson, P. J. (1991) Phosphatidylinositol membrane anchors and T-cell activation. *Immunol. Today* **12,** 35–41.
10. Rothberg, K. G., Heuser, J. E., Donzell, W. C., Ying, Y.-S., Glenney, J. R., and Anderson, R. G. W. (1992) Caveolin, a protein component of caveolae membrane coats. *Cell* **68,** 673–682.
11. Rothberg, K. G., Ying, Y.-S., Kolhouse, J. F., Kamen, B. A., and Anderson, R. G. W. (1990) The glycophospholipid-linked folate receptor internalizes folate without entering the clathrin-coated pit endocytic pathway. *J. Cell Biol.* **110,** 637–649.
12. Ying, Y., Anderson, R. G. W., and Rothberg, K. G. (1992) Each caveola contains multiple glycosyl-phosphatidylinositol anchored membrane proteins. *Cold Spring Harbor Symp. Quant. Biol.* **57,** 593–603.

13. Ferguson, M. A. J., Homans, S. W., Dwek, R. A., and Rademacher, T. W. (1988) Glycosyl-phosphatidylinositol moiety that anchors *Trypanosoma brucei* variant surface glycoprotein to the membrane. *Science* **239,** 753–759.
14. Griffiths, G. (1993) *Fine Structure Immunochemistry*. Springer-Verlag, Heidelberg. p. 459.
15. Mayor, S., Rothberg, K. G., and Maxfield, F. R. (1994) Sequestration of GPI-anchored proteins in caveolae triggered by cross-linking. *Science* **264,** 1948–1951.
16. Mayor, S. and Maxfield, F. R. (1995) Insolubility and redistribution of GPI-anchored proteins at the cell surface after detergent treatment. Mol. Biol. *Cell* **6,** 929–944.
17. Parton, R. G., Joggerst, B., and Simons, K. (1994) Regulated internalization of caveolae. *J. Cell Biol.* **127,** 1199–1215.
18. Coney, L. R., Tomassetti, A., Carayannopoulos, L., Frasca, V., Kamen, B. A., Colnaghi, M.I., and Zurawski, V. J. (1991) Cloning of a tumor-associated antigen: MOv18 and MOv19 antibodies recognize a folate-binding protein. *Cancer Res.* **51,** 6125–6132.
19. Howard, E. and Lane, D. (1988) *Antibodies: A Laboratory Manual*. Cold Spring Harbor Laboratory Press, Cold Spring Harbor, NY, p. 628.
20. Mayor, S., Presley, J. P, and Maxfield, F. R. (1993) Sorting of membrane components from endosomes and subsequent recycling to the cell surface occurs by a bulk flow process. *J. Cell Biol.* **121,** 1257–1269.
21. Hayat, M. (1981) *Principles and Techniques of Electron Microscopy*, Vol I, *Biological Applications*. Van Nostrand Reinhold, New York.
22. Davitz, M. A., Low, M. G., and Nussenzweig, V. (1986) Release of decay-accelerating factor (DAF) from the cell membrane by phosphatidylinositol-specific phospholipase C (PIPLC): selective modification of a complement regulatory protein. *J. Exp. Med.* **163,** 1150–1161.

4

Imaging Fluorescence Resonance Energy Transfer as Probe of Membrane Organization and Molecular Associations of GPI-Anchored Proteins

Anne K. Kenworthy and Michael Edidin

1. Introduction

1.1. Background

The spatial organization of (GPI)-anchored proteins in cell membranes is a matter of considerable interest. These proteins are thought to be organized into membrane microdomains enriched in GPI-anchored proteins, glycosphingolipids, cholesterol, and some other lipid-modified proteins. Such microdomains have been implicated in membrane trafficking and cell signaling events (reviewed in **ref. *1***). However, most evidence for the existence of microdomains comes from biochemical studies of isolated membrane fractions ***(1,2)***. Microscopy of intact cells has not detected microdomains enriched in GPI-anchored proteins ***(3–5)***; however, these experiments either sample a limited part of the cell surface at high electron-microscope resolution, or an entire cell at low light-microscope resolution.

To examine the organization of GPI-anchored proteins in intact cell membranes with high resolution, the authors have recently applied a novel method, imaging fluorescence resonance energy transfer (imaging FRET) ***(6)***. Because FRET occurs over distances of 1–10 nm, it is a useful tool for investigating molecular associations at a length-scale comparable to the size of the molecules themselves. When combined with fluorescence microscopy, FRET effectively increases the resolution of light microscopy to the molecular level, allowing one to perform imaging biochemistry. This chapter describes a method for performing imaging FRET measurements.

From: *Methods in Molecular Biology, Vol. 116: Protein Lipidation Protocols*
Edited by: M. H. Gelb © Humana Press Inc., Totowa, NJ

1.2. Imaging FRET

FRET involves the nonradiative transfer of energy from an excited state donor to a nearby acceptor (reviewed in **refs. *7*** and ***8***). The choice of fluorophores for FRET measurements requires that the emission spectrum of the donor overlap with the excitation spectrum of the acceptor. The energy transfer efficiency, *E*, is dependent on the inverse sixth power of the distance *r* between a donor and acceptor,

$$E = 1/\,(1 + (r/R_0)^6) \tag{1}$$

where R_0 is the characteristic Förster distance for the donor and acceptor fluorophores; at $r = R_0$, $E = 0.5$.

Experimentally, FRET can be detected from quenching of donor fluorescence, sensitized acceptor fluorescence, reduced donor lifetimes, and decreased rate of donor photobleaching. Imaging FRET measurements have been developed to measure all of these phenomena (reviewed in **refs. *7–12***).

The authors have recently developed a variation of imaging FRET, acceptor photobleaching imaging FRET ***(6)***. The method originally required a laser scanning confocal microscope ***(13–15)***, but it has been modified for use with a conventional fluorescence microscope ***(6)***. In this method, samples are double labeled with donor- and acceptor-conjugated antibodies against the molecules of interest (**Fig. 1**). If the donors and acceptors are in proximity, so that FRET can occur, then, in the presence of the acceptor, the donor fluorescence is quenched. After irreversible photobleaching of the acceptor, the donor fluorescence is no longer quenched by the acceptor, and donor fluorescence increases. Because the extent of steady state donor fluorescence quenching is directly related to the efficiency of energy transfer, FRET efficiencies can be calculated by

$$E = (I_{\mathrm{DA}\dagger} - I_{\mathrm{DA}})\,/\,I_{\mathrm{DA}\dagger} \tag{2}$$

where I_{DA} and $I_{\mathrm{DA}\dagger}$ are the donor fluorescence intensities before and after acceptor photobleaching, respectively. (It is also possible to measure energy transfer efficiencies from donor photobleaching kinetics ***[15]*** [*see* **refs. *16*** and ***17*** for more information].) An important feature of this method is that all the parameters needed to quantitate energy transfer efficiencies can be obtained from the same microscope field.

The principle of the acceptor photobleaching FRET method requires a photostable donor that is not detectably photobleached during the collection of two images, one before and one after bleaching the acceptor (**Fig. 1**). It also requires a photolabile acceptor, which can be completely photobleached in a relatively short time (**Fig. 1**). Two potential donor and acceptor pairs were surveyed for use in these experiments: fluorescein–sulforhodamine (Texas Red) and Cy3/

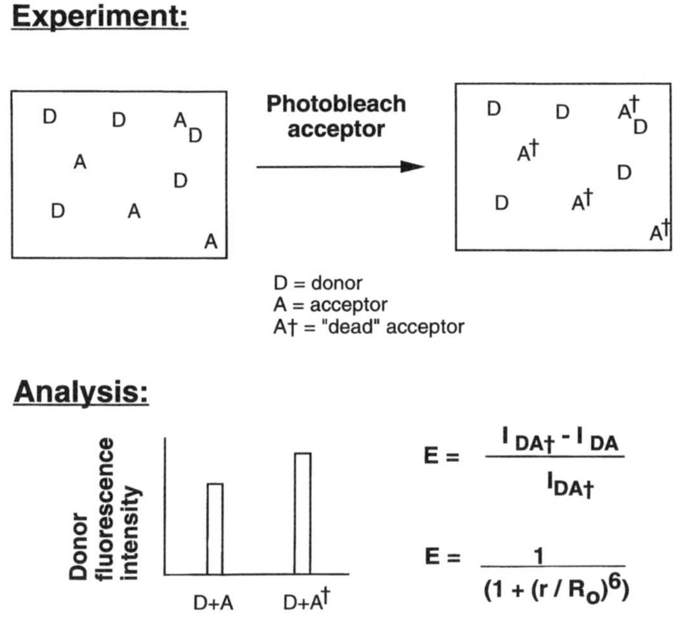

Fig. 1. Summary of the acceptor photobleaching FRET method.

Cy5 (**Fig. 2**). The photostability–photolability criteria were satisified by the donor and acceptor pair Cy3 (absorbance maximum 550 nm, emission maximum 570 nm) and Cy5 (absorbance maximum 649 nm, emission maximum 670 nm), two cyanine-based dyes *(18)* for which an R_0 of 50 Å has been calculated *(13)*. Cy5 (the acceptor) can be bleached completely by 3–5 min of continuous xenon arc lamp excitation; Cy3 (the donor) is more stable (**Fig. 2**). Cy3 and Cy5 can be directly conjugated to antibodies against the proteins of interest, and commercial secondary antibodies labeled with Cy3 and Cy5 are also available.

The use of a digital imaging system to perform these imaging FRET measurements is required for several reasons. First, Cy5 fluoresces in the far red, and is not detected well by the naked eye, but is readily detected by CCD cameras and photomultipliers (PMT). (Below, it is assumed that a CCD camera is used.) Second, the acceptor photobleaching FRET approach is sufficiently quantitative to allow for calculated E images to be generated, requiring a digital system (*see* **Subheading 3.3.**). Third, the quantitative information in the digitized donor and acceptor fluorescence intensities may be useful in interpreting the results (*see* **Subheading 3.4.**). Some relevant properties of digital imaging systems are reviewed in **ref. *9***.

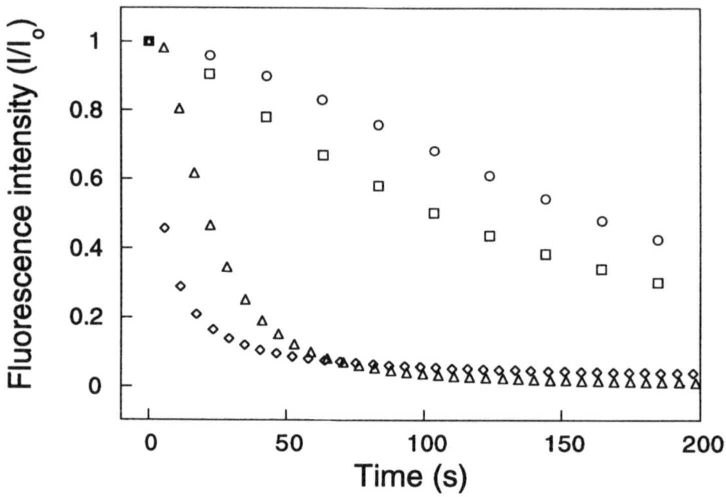

Fig. 2. Example of fluorophore photobleaching rates for two donor and acceptor pairs (Cy3/Cy5 and fluorescein/Texas Red) during continuous excitation with a xenon arc lamp. Cy3 (square) and Cy5 (triangle) were directly conjugated to a monoclonal antibody against 5' nucleotidase, as described in **Subheading 2.2.**, and fluorescein (diamond)- and Texas Red (circle)-conjugated antibodies were prepared using previously reported methods ***(19)***. The fluorophore-conjugated antibodies were used to label 5' nucleotidase transfected MDCK cells grown on glass cover slips ***(6)***. The samples were fixed, mounted in PBS on a slide, sealed with nail polish, and examined with filter sets optimized for each fluorophore. Mean fluorescence intensities over a 40 × 40 pixel region of interest from an individual cell were determined as a function of time. The fluorescence intensities at each time point (*I*) were normalized by the initial intensity (*I0*) of each sample.

2. Materials

2.1. Hardware and Software Requirements

1. Conventional or confocal fluorescence microscope with filter sets for imaging Cy3 and Cy5 fluorescence (i.e., Chroma Technologies, Brattleboro, VT; Omega Optical, Brattleboro, VT) coupled to a CCD camera or PMT.
2. Imaging software capable of performing image arithmetic and image registration.

2.2. Antibody Conjugation

1. Kits for directly conjugating Cy3 and Cy5 NHS esters to the protein of interest (Fluorolink™ labeling kits, Amersham, Arlington Heights, IL).
   a. Complete kits (including columns for separating free dye and conjugate).
   b. Dye packs only.

2. Column to separate free dye and conjugated protein, if dye packs only are purchased (i.e., Sephadex® G-50 or Bio-Gel® P-10; the authors have used Econo-Pac 10 DG disposable chromatography columns [Bio-Rad, Hercules, CA]).
3. Phosphate buffered saline (PBS; for use as a separation buffer; supplement with 0.1 % sodium azide, if desired).
4. 0.1 *M* sodium carbonate buffer, pH 9.3 (conjugation buffer).
5. Primary antibodies to be labeled (preferably at a concentration of ~1 mg/mL and a volume of 0.1 mL or 1 mL depending on the kit used).

3. Methods

3.1. Preparation of Fluorophore-Conjugated Antibodies

1. Label the antibodies according to manufacturer's instructions (the authors have labeled Fab, IgG, and polyclonal serum successfully with these kits). In brief:
   a. Exchange the antibody into 0.1 *M* carbonate buffer, pH 9.3.
   b. Incubate with prepackaged dye at recommended concentration (~1 mg/mL protein) and volume (1 mL or 0.1 mL) for recommended time.
   c. Separate the conjugated protein from the free dye.
   d. Calculate the dye to protein ratio (d/p) and protein concentration.
2. Antibodies are stored at 4°C with 0.02% azide or can be stored at –20°C; the relative advantages of each have been discussed ***(19)***.

3.2. Defining Imaging Parameters for Data Acquisition

3.2.1. Establishing Acceptor Photobleaching Parameters

1. Prepare two samples, one labeled with the Cy3-conjugated antibody only, and the other with the Cy5-conjugated antibody only. The samples should be fixed and prepared as for immunofluorescence microscopy, except that the sample should be mounted in buffer, and not in an antifading mounting medium (*see* **Notes**).
2. The sample labeled with Cy3 only is used to determine whether two subsequent images of Cy3 fluorescence can be obtained without significant bleaching. If the fluorescence intensity is decreased in the second image compared to the first, a neutral density filter should be used to decrease the excitation intensity since this affects the rate of bleaching.
3. The sample labeled with Cy5 only is used to determine the minimum time required to completely bleach the Cy5. This can be determined by collecting images of the Cy5 fluorescence as a function of time (as in **Fig. 2**). Because of the rapid photobleaching of Cy5, each test bleach should be performed on a new area of the slide.
4. The sample labeled with Cy3 only is then bleached using the Cy5 filters for the same time, and the fluorescence intensity of the Cy3 compared before and after the mock Cy5 bleach. If, under these conditions, there is a decrease in donor fluorescence, this effect may be eliminated by using a long pass (>570 nm) filter in the excitation path during the Cy5 bleach.

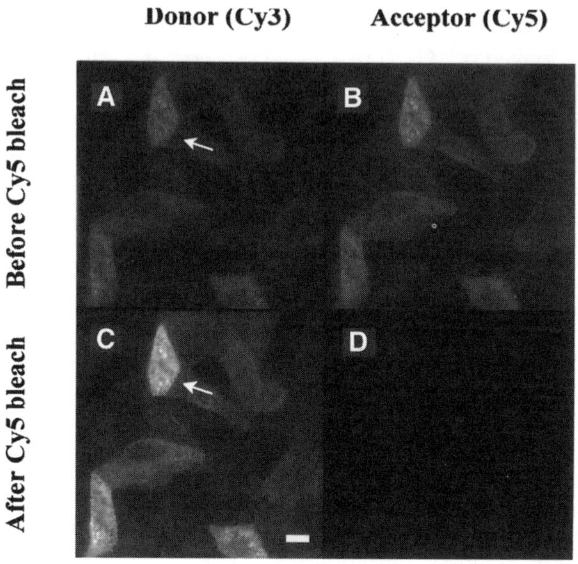

Fig. 3. Images collected during an acceptor photobleaching FRET experiment. Data are from a positive control experiment in which MDCK cells transfected with rat liver 5' NT ***(6)*** were labeled with 100 μg/mL Cy3-conjugated mouse monoclonal antibody against 5' NT, followed by 20 μg/mL Cy5-conjugated donkey antimouse IgG antibody, and then prepared for imaging as described in the legend to **Fig. 2**. **(A)** Cy3 (donor) before Cy5 (acceptor) photobleaching; **(B)** Cy5 before photobleaching; **(C)** Cy3 after Cy5 photobleaching; **(D)** Cy5 after photobleaching. Note the increased Cy3 fluorescence after Cy5 photobleaching (arrows in A and C). Bar, 10 μm.

5. Confirm that, using typical exposure times for image acquisition, no fluorescence is observed from a Cy3-labeled specimen when observed using the Cy5 filters, nor is Cy5 fluorescence detected using the Cy3 filter sets.

3.2.2. Data Acquisition

1. Prepare samples double labeled with the Cy3- and Cy5-conjugated antibodies against the proteins of interest (*see* **Subheading 3.4.** for considerations for labeling conditions.) Also prepare a sample labeled with Cy3 only, as a negative control for FRET. A useful positive control for FRET is to label samples with a donor-labeled primary antibody and an acceptor-labeled secondary antibody (or vice versa) (*see* **Fig. 3**).
2. If necessary, adjust the exposure time for Cy3 (donor) and Cy5 (acceptor) to maximize the full range of the CCD camera, keeping in mind that if FRET occurs, the donor fluorescence will increase after acceptor photobleaching.
3. Images should be collected in the following order for each field of cells (*see* **Fig. 3**):
   a. Cy3 (donor) prior to Cy5 (acceptor) photobleaching.
   b. Cy5 prior to photobleaching.

c. Cy5 at the completion of photobleaching.
d. Cy3 after Cy5 photobleaching.

Make sure to move to an area where the Cy5 has not been bleached before making a new measurement.

4. Measure the dark current contribution to the images by collecting an image with the shutter closed.

3.3. Calculation of Energy Transfer Efficiency Images

One distinct advantage of acceptor photobleaching FRET measurements over other imaging FRET methods is that images quantitatively mapping energy transfer efficiencies (E) can be easily calculated. E images are calculated using the images of donor fluorescence before (I_{DA}) and after ($I_{DA\dagger}$) acceptor photobleaching, from

$$E^{*}\mathbf{a} = [(I_{DA\dagger} - I_{DA}) / I_{DA\dagger}]^{*}\mathbf{b} \quad (3)$$

where **a** and **b** are scaling factors. The scaling factors are included because, depending on the software and the depth (i.e., 8- vs 12-bit) of the images, it is necessary to achieve a full range of E during this calculation, since, as written in **Eq. 2**, the resultant image would be reported in fractions. For example, for a 12-bit image, a scaling factor **b** of 10,000 is included in the division step. Thus, a 12-bit image is reported for E, ranging from 0 to 4096, so the scaling factor **a** is 100 when E is reported in %.

Before the E image calculation can be performed, it is necessary to subtract the dark current from each image; this constant noise contribution would, if included in the calculation, systematically decrease E. If appropriate, a correction can also be made for the autofluorescence of the cells. It is also necessary to register the donor images obtained before and after acceptor photobleaching, since subsequent images are typically randomly offset by a few pixels in either direction. Registration algorithims are available in some software packages, but this can also be done manually, if necessary, by translating the images slightly. Image registration is important, because, if the donor images are not correctly registered, artifacts such as systematic zeros will occur in the E image (**Fig. 4**). (It is often easier to detect problems in registration in the donor difference image, $[I_{DA\dagger} - I_{DA}]$, than in the final E image.) Differences in the focal plane of the donor images before and after acceptor photobleaching will also cause artifacts in the E image. These artifacts cannot be corrected, but they can be avoided (*see* above).

3.4. Biological Questions and Considerations for Experimental Design and Data Analysis

Below are issues relevant to experimental design and data analysis for several classes of questions that imaging FRET measurements can address.

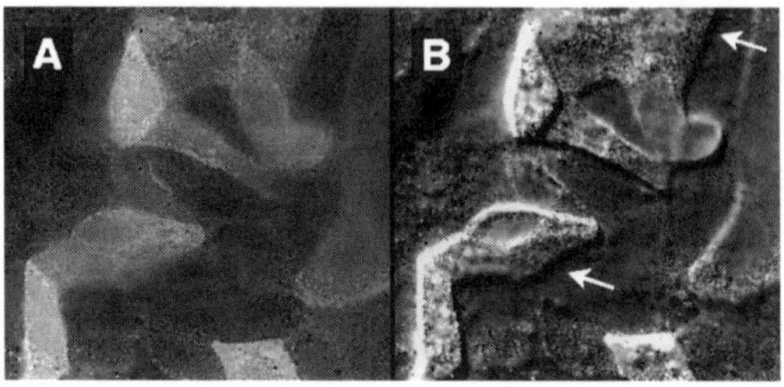

Fig. 4. Calculated *E* images for the data in **Fig. 3**, when registration of the donor images collected before and after acceptor photobleaching is either **(A)** correct or **(B)** incorrect. Notice the overall three dimensional appearance, and the areas where systematic zeros are observed (arrows in **B**), for the E image calculated from incorrectly registered donor images.

3.4.1. Is a Protein Randomly Distributed or Clustered?

The authors have recently described a method for investigating the lateral organization of a GPI-anchored protein on intact cell membranes, to determine whether the protein is randomly distributed or clustered in membrane microdomains ***(6)***. This method is an application of the theory for FRET on a surface ***(20,21)*** and relies on measurements of the dependence of FRET on both the donor and acceptor surface density (the number of fluorophores per unit area, which is proportional to the fluorescence intensity per unit area). For randomly distributed molecules, energy transfer efficiencies are a direct function of the surface density of acceptors, since the higher the surface density, the closer on average they will be to donors. Experimentally, the acceptor surface density can be changed in two ways: using a population of cells expressing different amounts of the proteins, or varying the ratio of donors and acceptors used to label the proteins of interest.

The dependence of E on donor and acceptor surface density can be evaluated from plots of E as a function of donor and acceptor fluorescence (**Fig. 5A**). The data shown in **Fig. 5A** are an example of proteins that are clustered; the acceptor-labeled secondary antibody is directly bound to the donor-labeled primary antibody. However, *E* is also dependent on acceptor surface density. This is because the protein recognized by the primary antibody is present at high surface density, so that acceptors on the secondary antibody can interact with donors in nearby primary/secondary antibody complex donors, as well as with donors on the antibodies to which the secondary is bound. In contrast, in the limit where clustered molecules are at low surface density, separated far enough so that the neighboring clusters do not interact with one another, *E* would be constant for all levels of acceptor fluorescence (assuming that the composition of the clusters is similar).

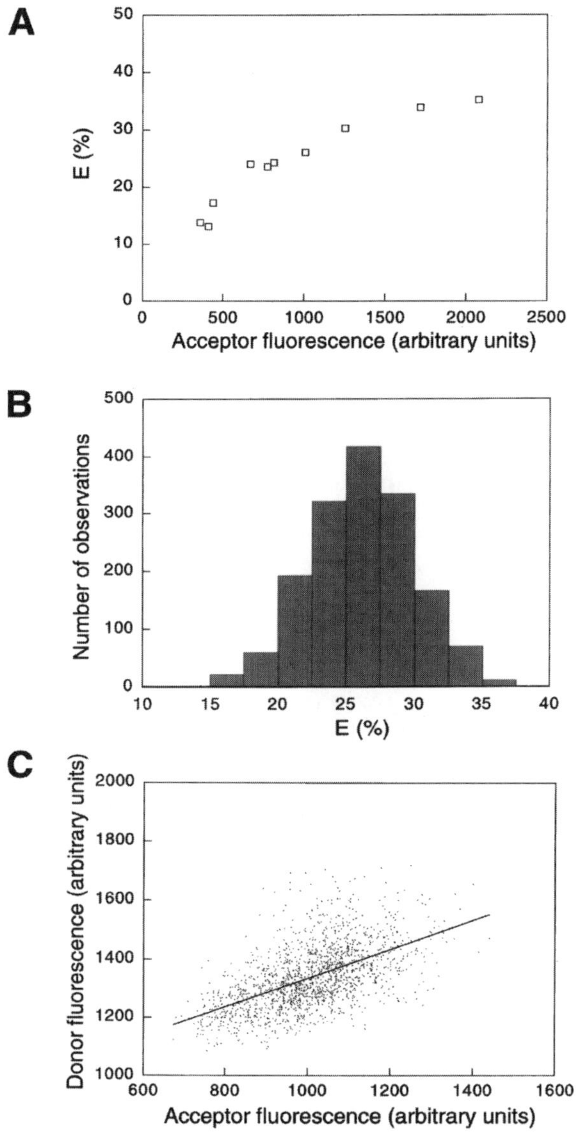

Fig. 5. Approaches for the analysis of FRET data. Data are from the experiment in **Fig. 3**. **(A)** E as a function of acceptor fluorescence intensity. Each datum shows the mean values for an individual cell, obtained for identical 40 × 40 pixel regions of interest from images of acceptor fluorescence intensities prior to photobleaching (**Fig. 3B**) and E (**Fig. 4A**). **(B)** Histogram of pixel-by-pixel variations in *E* for a single cell. Data were obtained on a pixel-by-pixel basis for a 40 × 40 pixel region of interest from an *E* image (**Fig. 4A**). **(C)** Relationship between donor and acceptor fluorescence. Data were obtained on a pixel-by-pixel basis for a single cell from identical 40 × 40 pixel regions of interest from images of acceptor fluorescence intensities prior to photobleaching (**Fig. 3B**) and donor fluorescence after acceptor photobleaching (**Fig. 3C**).

3.4.2. Does the Lateral Organization of a Protein Change?

FRET can also be used to test if the lateral organization of a protein changes as a function of time or treatment (for instance, during cell signaling). Changes in the lateral organization of a protein can be inferred by simply comparing *E* for the control and experimental cells. If the protein of interest is expressed at about the same level in both cell populations, then *E* can be compared using histograms of the distribution of *E* over a cell surface (**Fig. 5B**). A more quantitative analysis of these data can be used to model the structural changes reported by changes in *E* (for an example, *see* **ref. *22***).

3.4.3. Do Proteins, Co-localized by Conventional Microscopy, Interact at the Molecular Level?

Imaging FRET can be used to ask if molecules that are co-localized on the level of light microscopy (to within a few 100 nm of one another) are associated on the molecular level, since the lengthscales measured by FRET are on the order of the size of proteins themselves. (Studies of the lateral organization of a single kind of protein [*see* **Subheading 3.4.1.**] are an extreme example of this class of question.) For instance, imaging FRET has been used to map the α- and β-subunits of cholera toxin during its internalization and intracellular trafficking ***(14)*** and to probe the molecular proximity of proteins of the adherens junction ***(23)***. In each of these studies, in some cases, FRET was observed for co-localized proteins, and in other cases was not, demonstrating the resolving power of imaging FRET.

When possible, it may be useful to exchange the donor- and acceptor-labeled molecules in this kind of experiment, to confirm the specificity of FRET. Especially if one protein is much more abundant than the other, molecular proximities reported by FRET may be caused by a random, rather than specific, interaction. FRET would be significantly higher when the more abundant molecule is labeled with acceptor than when it is labeled with donor, because of the dependence of FRET between randomly distributed molecules on acceptor surface density.

It may be useful to compare the relative number of donor and acceptor fluorophores in a given pixel or region of interest (**Fig. 5C**). For such comparisons, donor fluorescence intensities after acceptor photobleaching, when donor fluorescence is fully dequenched, should be used. In the example shown, donor and acceptor fluorescence are highly correlated, because the donor-labeled molecule is binding directly to the acceptor-labeled molecule. For some applications, it may be useful to couple the number of donor and acceptor fluorophores in a given pixel with *E* for that pixel in the form of a two-dimensional histogram ***(14)***. This information can also be used to generate images that show structures mapping to the different populations revealed by the histogram ***(14)***.

4. Notes

1. Cells should be fixed and labeled using methods appropriate for the proteins of interest. The labeling conditions should be optimized to minimize nonspecific labeling, particularly if directly conjugated primary antibodies are used. A method for the preparation of Fab, and the relative merits of using a Fab vs an IgG for FRET measurements, have been recently discussed ***(19)***. If commercial Cy3- and Cy5-conjugated secondary antibodies are used (Amersham; Jackson ImmunoResearch Laboratories, West Grove, PA), control experiments should be performed to confirm that they do not induce clustering of the proteins being labeled ***(5)***. *See* **ref. *23*** for further discussion of the use of secondary antibodies in FRET experiments.
2. Cover slips are mounted on a slide using a buffer such as PBS as the mounting medium, and the cover slip is sealed with nail polish. The nail polish is allowed to dry for at least 1 h to stabilize the sample. This is important to help minimize changes in the focal plane of the image during an experiment, which can generate artifacts in the energy transfer efficiency image calculation. Samples are usually viewed within 24–48 h of preparation, since the buffer will evaporate unless thoroughly sealed.
3. The authors have observed two artifacts related to Cy5 photobleaching. First, for samples mounted in relatively small volumes of buffer, it is very difficult to bleach the Cy5. This might indicate the need for an excess of dissolved O_2 for photobleaching to occur, since, in the presence of antifading mounting medium (Vectashield, Vector Laboratories, Burlingame, CA), Cy5 is significantly more photostable than in PBS alone (A. Kenworthy and M. Edidin, unpublished observations). Second, the authors sometimes observe increased Cy5 fluorescence during continuous arc lamp excitation prior to complete photobleaching of the fluorophore. This could potentially be caused by the release of self-quenching of Cy5 conjugated to antibodies at high d/p ratios.
4. While acceptor photobleaching FRET allows one to quantitatively measure E, note that the absolute magnitude of E will depend on a number of factors in addition to the distance r between the donor- and acceptor-labeled antibodies against the proteins of interest (**Eq. 1**), including the ratio of donor- and acceptor-conjugated antibodies (**Subheading 3.4.**), the d/p ratios on the antibodies, and the relative orientations of the donor and acceptor fluorophores.
5. For their experiments, the authors use an integrated digital imaging system from Inovision Corporation (Research Triangle Park, NC); Inovision's Isee™ software is used both for data acquisition and image analysis.

Acknowledgments

The authors thank Thomas Jovin, Donna Ardnt-Jovin, and Philippe Bastiaens at the Max Planck Institute for Biophysical Chemistry in Göttingen, Germany, for introducing us to the acceptor photobleaching imaging FRET method. Work was performed at the Advanced Microscopy Facility in the Department of

Biology at Johns Hopkins University, and was supported by P01DK44375 from the National Institutes of Health to M. E.

References

1. Simons, K. and Ikonen, E. (1997) Functional rafts in cell membranes. *Nature* **387,** 569–572.
2. Brown, D. A. and Rose, J. K. (1992) Sorting of GPI-anchored proteins to glycolipid-enriched membrane subdomains during transport to the apical cell surface. *Cell* **68,** 533–544.
3. Fujimoto, T. (1996) GPI-anchored proteins, glycosphingolipids, and sphingomyelin are sequestered to caveolae only after crosslinking. *J. Histochem. Cytochem.* **44,** 929–941.
4. Mayor, S. and Maxfield, F. R. (1995) Insolubility and redistribution of GPI-anchored proteins at the cell surface after detergent treatment. *Mol. Biol. Cell.* **6,** 929–944.
5. Mayor, S., Rothberg, K. G., and Maxfield, F. R. (1994) Sequestration of GPI-anchored proteins in caveolae triggered by cross-linking. *Science* **264,** 1948–1951.
6. Kenworthy, A. K. and Edidin, M. (1997) Distribution of a glycosylphosphatidylinositol-anchored protein at the apical surface of MDCK cells examined at a resolution of <100Å using imaging fluourescence response energy transfer. *J. Cell Biol.* **142,** 69–84.
7. Clegg, R. M. (1995) Fluorescence resonance energy transfer. *Curr. Opin. Biotechnol.* **6,** 103–110.
8. Wu, P. and Brand, L. (1994) Resonance energy transfer: methods and applications. *Anal. Biochem.* **218,** 1–13.
9. Dunn, K. W., Mayor, S., Myers, J. N., and Maxfield, F. R. (1994) Applications of ratio fluorescence microscopy in the study of cell physiology. *FASEB J.* **8,** 573–582.
10. Herman, B. (1989) Resonance energy transfer microscopy. *Methods Cell Biol.* **30,** 219–243.
11. Selvin, P. R. (1995) Fluorescence resonance energy transfer. *Methods Enzymol.* **246,** 300–334.
12. Tsien, R. Y., Bacskai, B. J., and Adams, S. R. (1993) FRET for studying intracellular signalling. *Trends Cell Biol.* **3,** 242–245.
13. Bastiaens, P. I. and Jovin, T. M. (1996) Microspectroscopic imaging tracks the intracellular processing of a signal transduction protein: fluorescent-labeled protein kinase C beta I. *Proc. Natl. Acad. Sci. USA* **93,** 8407–8412.
14. Bastiaens, P. I., Majoul, I. V., Verveer, P. J., Soling, H. D., and Jovin, T. M. (1996) Imaging the intracellular trafficking and state of the AB5 quaternary structure of cholera toxin. *EMBO J.* **15,** 4246–4253.
15. Bastiaens, P. I. H., Wouters, F. S., and Jovin, T. M. (1995) Imaging the molecular state of proteins in cells by fluorescence resonance energy transfer (FRET). Sequential photobleaching of Förster donor-acceptor pairs, in *2nd Hamamatsu International Symposium on Biomolecular Mechanisms and Photonics: Cell-Cell Communications.*

16. Jovin, T. M. and Arndt-Jovin, D. J. (1989) FRET microscopy: digital imaging of fluorescence resonance energy transfer. Application in cell biology, in *Cell Structure and Function by Microspectrofluorimetry,* (Kohen, E., Ploem, J. S., and Hirschberg, J. G., eds.), Academic, Orlando, FL, pp. 99–117.
17. Jovin, T. M. and Arndt-Jovin, D. J. (1989) Luminescence digital imaging microscopy. *Annu. Rev. Biophys. Biophys. Chem.* **18,** 271–308.
18. Mujumdar, R. B., Ernst, L. A., Mujumdar, S. R., Lewis, C. J., and Waggoner, A. S. (1993) Cyanine dye labeling reagents: sulfoindocyanine succinimidyl esters. *Bioconjugate Chem.* **4,** 105–111.
19. Matko, J. and Edidin, M. (1997) Energy transfer methods for detecting molecular clusters on cell surfaces. *Methods Enzymol.* **278,** 444–462.
20. Dewey, T. G. and Hammes, G. G. (1980) Calculation of fluorescence resonance energy transfer on surfaces. *Biophys. J.* **32,** 1023–1035.
21. Wolber, P. K. and Hudson, B. S. (1979) An analytic solution to the Förster energy transfer problem in two dimensions. *Biophys. J.* **28,** 197–210.
22. Gadella, T. W., Jr., and Jovin, T. M. (1995) Oligomerization of epidermal growth factor receptors on A431 cells studied by time-resolved fluorescence imaging microscopy. A stereochemical model for tyrosine kinase receptor activation. *J.Cell Biol.* **129,** 1543–1558.
23. Kam, Z., Volberg, T., and Geiger, B. (1995) Mapping of adherens junction components using microscopic resonance energy transfer imaging. *J.Cell Sci.* **108,** 1051–1062.

5

Purification of Caveolae-Derived Membrane Microdomains Containing Lipid-Anchored Signaling Molecules, Such as GPI-Anchored Proteins, H-Ras, Src-Family Tyrosine Kinases, eNOS, and G-Protein α-, β-, and γ-Subunits

Michael P. Lisanti, Massimo Sargiacomo, and Philipp E. Scherer

1. Introduction

This laboratory is interested in understanding how signaling molecules are compartmentalized within normal cells, and how this molecular organization may be disrupted during oncogenic transformation. For this purpose, the authors have focused their efforts on a particular organelle, or subcompartment of the plasma membrane, known as caveolae *(1)*.

Caveolae are 50–100-nm vesicular structures that are located at or near the plasma membrane *(2,3)*. Membrane preparations enriched in purified caveolae contain lipid-modified signaling molecules (such as heterotrimeric G proteins, Ras, Src-family tyrosine kinases, and nitric oxide synthase [eNOS]) and caveolin, a transformation-dependent v-Src substrate *(4–17)*.

Caveolin, a 21–24 kDa integral membrane protein, is a principal component of caveolin in vivo *(18–20)*, and serves are a marker protein for this organelle *(4–6)*. Recent studies have shown that caveolin is only the first member of a growing gene family of caveolin proteins; caveolin has been retermed caveolin-1 *(21)*. Three different caveolin genes (Cav-1, Cav-2, and Cav-3), encoding four different subtypes of caveolin, have been described thus far *(22–25)*. The working hypothesis here is that caveolins act as scaffolding proteins to recruit and sequester diverse signaling molecules within caveolin-rich areas of the

From: *Methods in Molecular Biology, Vol. 116: Protein Lipidation Protocols*
Edited by: M. H. Gelb © Humana Press Inc., Totowa, NJ

plasma membrane for presentation to activated receptors in a ligand-dependent manner (the authors call this the "caveolae signaling hypothesis") ***(1,26–28)***.

This hypothesis has now been independently validated using several receptor systems for growth factor receptors (EGF-R, PDGF-R, Insulin-R) ***(9,12–14,29,30)*** and G-protein coupled receptors (endothelin, bradykinin, and mAch-R) ***(29,31–33)***. Here is presented two now very widely used methods for the purification of caveolae-derived membranes from cultured cells and whole tissues ***(4,5,10,34)***.

2. Experimental Strategies

Caveolae have been purified from cultured cells and whole tissue, based on their unusual biochemical and biophysical properties ***(4–6,9,10,34)***. Caveolae membranes are highly enriched in glyco-sphingolipids and cholesterol, making them extremely light, and, therefore, buoyant in sucrose density gradients. This unique lipid composition also confers resistance to solubilization by nonionic detergents such as Triton X-100 and NP-40 (at low temperatures) ***(35,36)***. For example, when intact cells were fixed in paraformaldehyde, extracted with Triton X-100, and then examined by electron microscopy, the insoluble membranes that remained were found to be caveolae ***(37)***.

These special properties, combined with the use of caveolins as marker proteins, allow the purification of caveolae from other cell organelles, as assessed using specific marker proteins for ER, Golgi, mitochondria, lysosomes, and noncaveolar plasma membranes ***(4,5,10,34)***.

In addition, a polyhistidine-tagged form of caveolin-1 has been recombinantly expressed and used to affinity-purify caveolae via a detergent-free scheme in mammalian cells ***(10)*** and in Sf 21 insect cells ***(38)***.

3. Methods

3.1. Triton-Based Purification of Caveolae-Derived Membranes

3.1.1. Materials and Equipment

1. MDCK cells (six 150-mm dishes at confluent density).
2. Phosphate buffered saline (PBS).
3. (2-[N-Morpholino]ethane sulfonic acid) (Mes)-buffered saline (25 m*M* Mes, pH 6.5, 0.15 *M* NaCl).
4. Mes-buffered saline, plus 1% Triton and PMSF (75 µL of saturated ice-cold ethanol solution per 10 mL of buffer).
5. 80% (w/v) sucrose prepared in Mes-buffered saline lacking detergent or PMSF.
6. Dounce homogenizer.
7. Pharmacia peristaltic pump (P-1) with attached gradient chamber.
8. Beckman SW41 rotor with corresponding ultraclear ultracentrifuge tubes.
9. Beckman L8 ultracentrifuge.

10. Liquid nitrogen.
11. –80°C Storage capacity.

3.1.2. Cell Culture

Serially transfer MDCK cells (or another cell line) by trypsinization from one T75 flask to three T75 flasks, and then to six 150 mm dishes. Let cells grow to confluence for at least 2 d before harvesting. Use DME medium (Gibco-BRL, cat. no. 11995-065) containing pen/strep and 5% FBS. Use one plate per gradient. Omit any antifungal agents such as Nystatin or Fungizone, because these are cholesterol-binding antibiotics that will disrupt caveolae architecture.

3.1.3. Cell Extraction and Fractionation

1. Rinse each 150-mm dish with ice-cold PBS (3X, 20 mL each). Thoroughly remove PBS, and add 1 mL of ice-cold Mes-buffered saline (25 m*M* Mes, pH 6.5, 0.15 *M* NaCl), plus 1% Triton X-100 and PMSF (75 µL of saturated ice-cold ethanol solution per 10 mL of buffer). Collect cells with a cell scraper, and place in an 15-mL tube on ice. Rinse dish with an additional 1 mL lysis buffer to collect remaining cells, and pool with first 1 mL.
2. Homogenize with 8–10 strokes of a Dounce homogenizer. Adjust to 40% sucrose by adding an equal volume (2 mL) of ice-cold 80% sucrose prepared in Mes-buffered saline lacking detergent or PMSF. Vortex briefly, until homogeneous.
3. Place homogenate at the bottom of an ultraclear Beckman ultracentrifuge tube and form a linear 5–30% gradient atop it at a flow rate, not more than 0.5 mL/min. Use solutions of 5% (4 mL) and 30% (4 mL) sucrose prepared in Mes-buffered saline lacking detergent and PMSF. Adjust the weight of each gradient to within 0.1 g.
4. Centrifuge at 200,000 *g* for 12–16 h (overnight) at 4°C in a Beckman SW41 rotor.
5. Collect 13 1-mL fractions from the top. Fraction 13 is the insoluble pellet. Snap freeze in liquid nitrogen and store frozen at –80°C.
6. Alternatively, the opaque band (migrating at ~10–20% sucrose) is harvested, diluted with Mes-buffered saline lacking detergent or PMSF and pelleted in the microcentrifuge. Pellets are snap-frozen in liquid nitrogen and stored frozen at –80°C.
7. A 150-mm dish of MDCK cells, representing 9–11 mg protein, yields ~4–6 µg of caveolae-enriched membrane domains, i.e., ~0.05% of the initial homogenate. Triton extraction solubilizes ~85% of the protein (8.5 mg), the majority of which remains in the 40% sucrose layer, ~1.5 mg forms an insoluble pellet below the 40% sucrose.
8. Organelle-specific membrane marker assays can be performed as described ***(39,40)***. These assay systems are not affected by the presence or absence of 1% Triton X-100 in the initial homogenate (*see* **Note 1**).

3.2. Detergent-Free Purification of Caveolae-Derived Membranes

3.2.1. Materials and Equipment

1. MDCK cells (six 150-mm dishes at confluent density).
2. PBS.
3. 500 m*M* sodium carbonate, pH 11.
4. Mes-buffered saline (25 m*M* Mes, pH 6.5, 0.15 *M* NaCl).
5. Mes-buffered saline plus 1% Triton and PMSF (75 µL of saturated ice-cold ethanol solution per 10 mL of buffer).
6. 90% (w/v) Sucrose prepared in Mes-buffered saline lacking detergent or PMSF.
7. Polytron tissue grinder (Kinematica GmbH from Brinkmann Instruments, Westbury, NY).
8. Branson Sonifier 250 (Branson Ultrasonic Corp., Danbury, CT).
9. Beckman SW41 rotor with corresponding ultraclear ultracentrifuge tubes.
10. Beckman L8 ultracentrifuge.
11. Liquid nitrogen.
12. –80°C storage capacity.

3.2.2. Cell Culture

As described above for Triton X-100 based purification of caveolae membranes.

3.2.3. Cell Extraction and Fractionation

1. Wash cells three times with ice-cold PBS
2. Scrape cells into 2 mL total of ice-cold 500 m*M* sodium carbonate.
3. Place cells on ice. Polytron at maximum setting for 30 s.
4. Next, sonicate cells constantly for 30 s at setting 2 on the sonicator. Next, sonicate cells for 30-s bursts at a setting of 4 or 5. Look at the lysate. If it is relatively clear and free of clumps, then it is ready to use. If not, sonicate an additional 30 s at setting 2. **Note:** All is done on ice.
5. Add the 2 mL of cell lysate to 2 mL of 90% sucrose in a 14 × 89 mm Beckman centrifuge ultraclear tube (the volume of tube is a little more than 12 mL). Vortex until homogeneous. Note the final concentration of sucrose is now 45%.
6. Atop the 45% layer, add 4 mL of 35% sucrose and 4 mL of 5% sucrose, respectively, to generate a discontinuous sucrose gradient. Be careful when adding respective layers, to avoid disrupting the layer beneath it.
7. Centrifuge at 200,000*g* for 12–16 h (overnight) in a Beckman SW41 rotor.
8. Collect 13 1-mL fractions from the top. Fraction 13 is the insoluble pellet. Snap freeze in liquid nitrogen and store frozen at –80°C. A light-scattering band confined to the 5–35% sucrose interface is observed that contains caveolin, but excludes most other cellular proteins.
9. Analyze by SDS-PAGE.

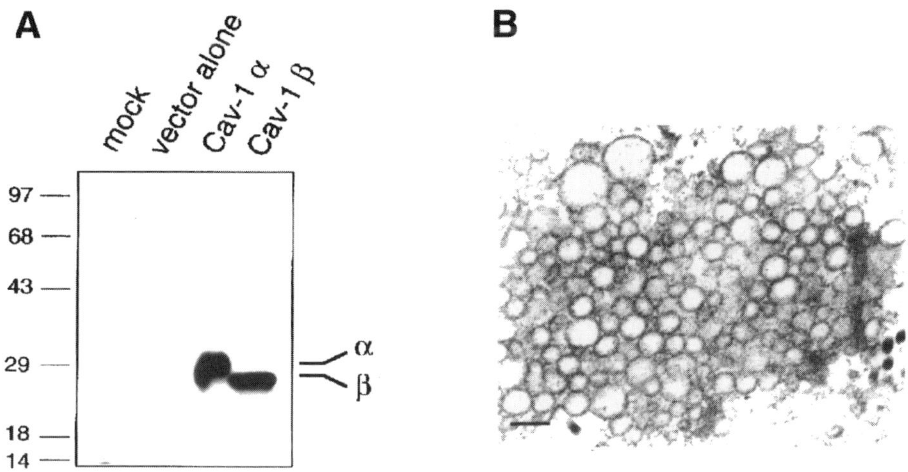

Fig. 1. Purification of "recombinant caveolae" from Sf 21 insect cells infected with a baculovirus-based vector harboring caveolin-1. **(A)** Expression of caveolin-1α and caveolin-1β in Sf 21 insect cells by infection with recombinant baculoviruses. **(B)** Morphology of caveolin-1 infected Sf 21 insect cells. The entire cytoplasm is filled with a homogenous population of caveolae-sized vesicles. Bar = 100 nm. *(continued)*

3.3. New Directions

Methods have recently been developed for the generation and purification of recombinant caveolae through the overexpression of caveolin in insect cells (**Fig. 1**; *[38]*). When Sf21 insect cells are infected with a recombinant baculovirus encoding the caveolin-1 protein, they accumulate hundreds of 50- to 120-nm (diameter) cytoplasmic vesicles. These cytoplasmic vesicles are the same size as caveolae and plasmalemmal vesicles seen in mammalian cells. The abundance of these cytoplasmic vesicles and their ease of purification make for an attractive method to generate large amounts of recombinant caveolae. Furthermore, co-expression of caveolin-1 and other proteins is possible with this system.

4. Notes and Comments

1. For the preparation of 5% and 30% sucrose solutions, dilute the 80% stock with Mes-buffered saline lacking detergent or PMSF. Dilute 37.5 mL of 80% sucrose to 100 mL to yield 30% sucrose. Dilute 6.25 mL of 80% sucrose to 100 mL to yield 5% sucrose.
2. 90% sucrose is prepared by using 90% wt/vol of sucrose in Mes buffered saline. Final concentration of Mes is 10 m*M*, pH6.5, and that of NaCl is 150 m*M*. The rest of the volume is water. Since 90% sucrose is hard to dissolve, it is generally heated. First, heat a beaker of water until it boils. Next, place the 90% sucrose

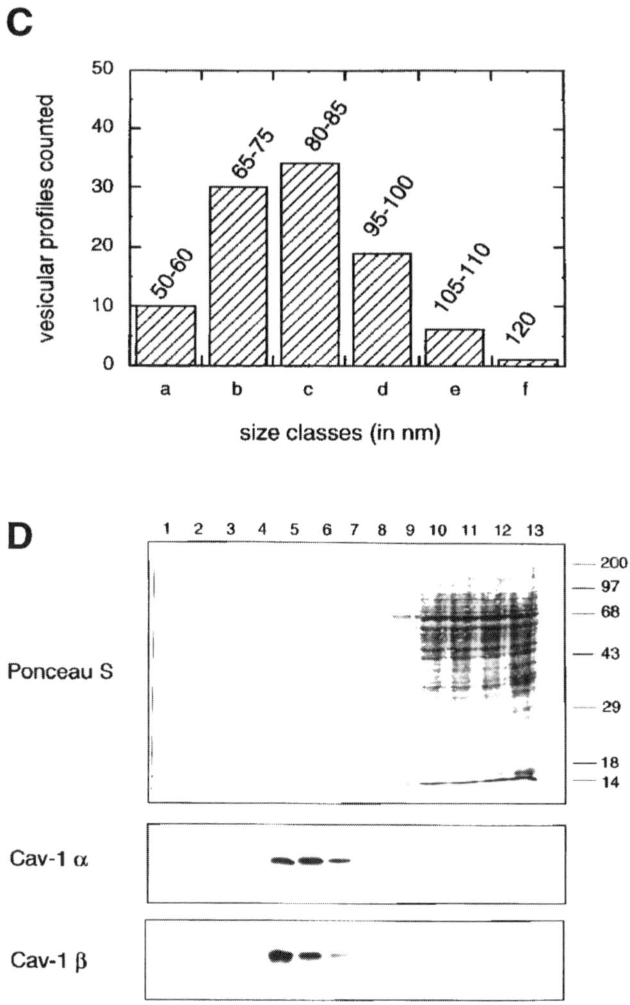

Fig. 1. *(continued)* **(C)** Quantitative morphological analysis indicates that ~68% of these caveolin induced vesicles have a diameter of 65–95 nm (80.3 ± 14.8 nm). **(D)** Purification of caveolin-induced vesicles using Method 2 as described above (*see* **Subhead 3.2.**).

mixture, which is in a smaller beaker/bottle, in the boiling water. Stir the sucrose until it completely dissolves. Store 90% sucrose at room temperature because it crystallizes at colder temperatures.

35 and 5% sucrose are prepared by diluting the 90% sucrose into MBS containing 250 m*M* sodium carbonate. Store these solutions at 4°C.

E

F

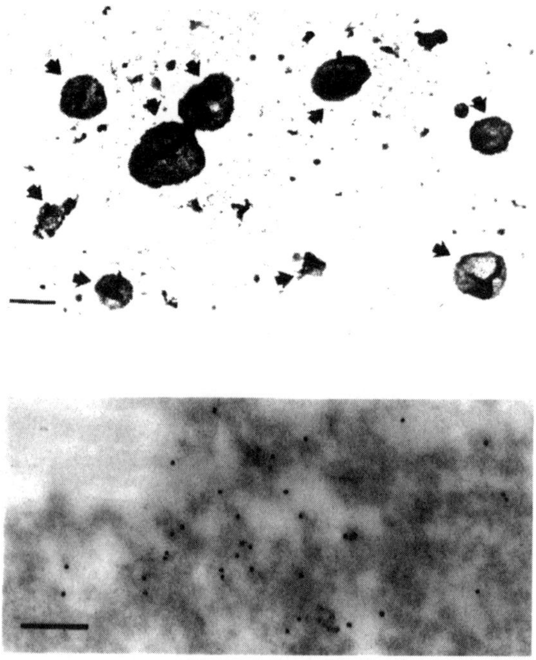

Fig. 1. *(continued)* **(E)** Morphology of purified caveolin-induced vesicles as seen by whole-mount electron microscopy. This procedure, which preserves membrane structure, stains membranes black. These purified recombinant caveolae appeared as ~50–100 nm vesicular structures (arrowheads) and were indistinguishable from preparations of native mammalian caveolae visualized by the same technique. Bar = 100 nm. **(F)** Immunolabeling of purified caveolin-induced vesicles. Vesicles were sedimented and processed as a packed pellet for fixation in paraformaldehyde and immuno-gold labeling with anti-caveolin-1 IgG (mAb 2297). Bar = 100 nm. (Adapted with permission from **ref. 38**.)

A stock solution of 0.5 *M* sodium carbonate is made by simply dissolving sodium carbonate in water. The pH of the solution is about 11. The pH does not need to be adjusted.

Acknowledgments

The authors thank members of the Scherer and Lisanti laboratories for helpful and insightful discussions, and John R. Glenney and Roberto Campos-Gonzalez at Transduction Laboratories (Lexington, KY) for generously donating anti-caveolin-1 antibodies. This work was supported by an NIH grant from the NCI R01-CA-80250 (to M. P. L.), and grants from the Charles E.

Culpeper Foundation (to M. P. L.), G. Harold and Leila Y. Mathers Charitable Foundation (to M. P. L. and P. E. S.) and the Sidney Kimmel Foundation for Cancer Research (to M. P. L.). P. E. S. was supported by a grant from Pfizer Corp., a pilor grant from the AECOM Diabetes Research & Training Center (DRTC), and by a research grant from the American Diabetes Association.

References

1. Lisanti, M. P., Scherer, P., Tang, Z.-L., and Sargiacomo, M. (1994) Caveolae, caveolin and caveolin-rich membrane domains: a signalling hypothesis. *Trends Cell Biol.* **4,** 231–235.
2. Engelman, J. A., Zhang, X. L., Galbiati, F., Volonte, D., Sotgia, F., Pestell, R. G., Minetti, C., Scherer, P. E., Okamoto, T., and Lisanti, M. P. (1998) Molecular genetics of the caveolin gene family: implications for human cancers, diabetes, Alzheimer's disease, and muscular dystrophy. *Am. J. Hum. Genetics* **63,** 1578–1587.
3. Okamoto, T., Schlegel, A., Scherer, P. E., and Lisanti, M. P. (1998) Caveolins, a family of scaffolding proteins for organizing "pre-assembled signaling complexes" at the plasma membrane. *J. Biol. Chem.* **273,** 5419–5422.
4. Sargiacomo, M., Sudol, M., Tang, Z.-L., and Lisanti, M. P. (1993) Signal transducing molecules and GPI-linked proteins form a caveolin-rich insoluble complex in MDCK cells. *J. Cell Biol.* **122,** 789–807.
5. Lisanti, M. P., Scherer, P.E., Vidugiriene, J., Tang, Z-L., Hermanoski-Vosatka, A., Tu, Y.-H., Cook, R. F., and Sargiacomo, M. (1994) Characterization of caveolin-rich membrane domains isolated from an endothelial-rich source: implications for human disease. *J. Cell Biol.* **126,** 111–126.
6. Chang, W. J., Ying, Y., Rothberg, K., Hooper, N., Turner, A., Gambliel, H., et al. (1994) Purification and characterization of smooth muscle cell caveolae. *J. Cell Biol.* **126,** 127–138.
7. Shenoy-Scaria, A. M., Dietzen, D. J., Kwong, J., Link, D. C., and Lublin, D. M. (1994) Cysteine-3 of Src family tyrosine kinases determines palmitoylation and localization in caveolae. *J. Cell Biol.* **126,** 353–363.
8. Robbins, S. M., Quintrell, N. A., and Bishop, M. J. (1995) Myryistoylation and differential palmitoylation of the HCK protein tyrosine kinases govern their attachment to membranes and association with caveolae. *Mol. Cell. Biol.* **15,** 3507–3515.
9. Smart, E. J., Ying, Y., Mineo, C., and Anderson, R. G. W. (1995) A detergent free method for purifying caveolae membrane from tissue cultured cells. *Proc. Natl. Acad. Sci. USA* **92,** 10104–10108.
10. Song, K. S., Li, S., Okamoto, T., Quilliam, L., Sargiacomo, M., and Lisanti, M. P. (1996) Copurification and direct interaction of Ras with caveolin, an integral membrane protein of caveolae microdomains. Detergent free purification of caveolae membranes. *J. Biol. Chem.* **271,** 9690–9697.
11. Schnitzer, J. E., Liu, J., and Oh, P. (1995) Endothelial caveolae have the molecular transport machinery for vesicle budding, docking, and fusion including VAMP, NSF, SNAP, annexins, and GTPases. *J. Biol. Chem.* **270,** 14,399–14,404.

12. Liu, P., Ying, Y., Ko, Y.-G., and Anderson, R. G. W. (1996) Localization of the PDGF-stimulated phosphorylation cascade to caveolae. *J. Biol. Chem.* **271,** 10,299–10,303.
13. Liu, J., Oh, P., Horner, T., Rogers, R. A., and Schnitzer, J. E. (1997) Organized endothelial cell signal transduction in caveolae. *J. Biol. Chem.* **272,** 7211–7222.
14. Mineo, C., James, G. L., Smart, E. J., and Anderson, R. G. W. (1996) Localization of the EGF-stimulated Ras/Raf-1 interaction to caveolae membrane. *J. Biol. Chem.* **271,** 11,930–11,935.
15. Garcia-Cardena, G., Fan, R., Stern, D., Liu, J., and Sessa, W. (1996) Endothelial nitric oxide synthase is regulated by tyrsosine phosphorylation and interacts with caveolin-1. *J. Biol. Chem.* **271,** 27,237–27,240.
16. Shaul, P. W., Smart, E. J., Robinson, L. J., German, Z., Yuhanna, I. S., Ying, Y., Anderson, R. G. W., and Michel, T. (1996) Acylation targets endothelial nitric-oxide synthase to plasmalemmal caveolae. *J. Biol. Chem.* **271,** 6518–6522.
17. Feron, O., Belhhassen, L., Kobzik, L., Smith, T. W., Kelly, R. A., and Michel, T. (1996) Endothelial nitric oxide synthase targeting to caveolae. Specific interactions with caveolin isoforms in cardiac myocytes and endothelial cells. *J. Biol. Chem.* **271,** 22,810–22,814.
18. Glenney, J. R. and Soppet, D. (1992) Sequence and expression of caveolin, a protein component of caveolae plasma membrane domains phosphorylated on tyrosine in RSV-transformed fibroblasts. *Proc. Natl. Acad. Sci. USA* **89,** 10,517–10,521.
19. Glenney, J. R. (1992) The sequence of human caveolin reveals identity with VIP 21, a component of transport vesicles. *FEBS Lett.* **314,** 45–48.
20. Rothberg, K. G., Heuser, J. E., Donzell, W. C., Ying, Y., Glenney, J. R., and Anderson, R. G. W. (1992) Caveolin, a protein component of caveolae membrane coats. *Cell.* **68,** 673–682.
21. Couet, J., Li, S., Okamoto, T., Scherer, P. S., and Lisanti, M. P. (1997) Molecular and cellular biology of caveolae: pardoxes and plasticities. *Trends Cardiovasc. Med.* **7,** 103–110.
22. Scherer, P. E., Tang, Z.-L., Chun, M. C., Sargiacomo, M., Lodish, H. F., and Lisanti, M. P. 1995. Caveolin isoforms differ in their N-terminal protein sequence and subcellular distribution: Identification and epitope mapping of an isoform-specific monoclonal antibody probe. *J. Biol. Chem.* **270,** 16,395–16,401.
23. Scherer, P. E., Okamoto, T., Chun, M., Nishimoto, I., Lodish, H. F., and Lisanti, M. P. (1996) Identification, sequence and expression of caveolin-2 defines a caveolin gene family. *Proc. Natl. Acad. Sci. USA* **93,** 131–135.
24. Tang, Z.-L., Scherer, P. E., Okamoto, T., Song, K., Chu, C., Kohtz, D.S., et al. (1996) Molecular cloning of caveolin-3, a novel member of the caveolin gene family expressed predominantly in muscle. *J. Biol. Chem.* **271,** 2255–2261.
25. Way, M. and Parton, R. (1995) M-caveolin, a muscle-specific caveolin-related protein. *FEBS Lett.* **376,** 108–112.
26. Sargiacomo, M., Scherer, P. E., Tang, Z.-L., Kubler, E., Song, K. S., Sanders, M. C., and Lisanti, M. P. (1995) Oligomeric structure of caveolin: implications for caveolae membrane organization. *Proc. Natl. Acad. Sci. USA* **92,** 9407–9411.

27. Li, S., Couet, J., and Lisanti, M. P. (1996) Src tyrosine kinases, G alpha subunits and H-Ras share a common membrane-anchored scaffolding protein, caveolin. Caveolin binding negatively regulates the auto-activation of Src tyrosine kinases. *J. Biol. Chem.* **271,** 29,182–29,190.
28. Couet, J., Li, S., Okamoto, T., Ikezu, T., and Lisanti, M. P. (1997) Identification of peptide and protein ligands for the caveolin-scaffolding domain. Implications for the interaction of caveolin with caveolae-associated proteins. *J. Biol. Chem.* **272,** 6525–6533.
29. Pike, L. and Casey, L. (1996) Localization and turnover of phosphatidylinositol 4,5-bisphosphate in caveolin-enriched membrane domains. *J. Biol. Chem.* **271,** 26,453–26,456.
30. Corley-Mastick, C., Brady, M. J., and Saltiel, A. R. (1995) Insulin stimulates the tyrosine phosphorylation of caveolin. *J. Cell Biol.* **129,** 1523–1531.
31. Chun, M., Liyanage, U., Lisanti, M. P., and Lodish, H. F. (1994) Signal transduction of a G-protein coupled receptor in caveolae: colocalization of endothelin and its receptor with caveolin. *Proc. Natl. Acad. Sci. USA* **91,** 11,728–11,732.
32. de Weerd, W. and Leeb-Lundberg, L. (1997) Bradykinin sequesters B2 bradykinin receptors and the receptor-coupled galpha subunits G alpha q and G alpha i in caveolae in DDT1 MF-2 smooth muscle cells. *J. Biol. Chem.* **272,** 17,858–17,866.
33. Feron, O., Smith, T., Michel, T., and Kelly, R. (1997) Dynamic targeting of the agonist-stimulated m2 muscarinic acetylcholine receptor to caveolae in cardiac myocytes. *J. Biol. Chem.* **272,** 17,744–17,748.
34. Scherer, P. E., Lisanti, M. P., Baldini, G., Sargiacomo, M., Corley-Mastick, C., and Lodish, H. F. (1994) Induction of caveolin during adipogenesis and association of GLUT4 with caveolin-rich vesicles. *J. Cell Biol.* **127,** 1233–1243.
35. Schroeder, R., London, E., and Brown, D. (1994) Interactions between saturated acyl chains confer detergent resistance on lipids and GPI-anchored proteins: GPI-anchored proteins in liposomes and cells show similar behavior. *Proc. Natl. Acad. Sci. USA* **91,** 12,130–12,134.
36. Ferraretto, A., Pitto, M., Palestini, P., and Masserini, M. (1997) Lipid domains in the membrane: thermotropic properties of sphingomyelin vesicles containing GM1 ganglioside and cholesterol. *Biochemistry* **36,** 9232–9236.
37. Moldovan, N., Heltianu, C., Simionescu, N., and Simionescu, M. (1995) Ultrastructural evidence of differential solubility in Triton X-100 of endothelial vesicles and plasma membrane. *Exp. Cell Res.* **219,** 309–313.
38. Li, S., Song, K. S., Koh, S., and Lisanti, M. P. (1996) Baculovirus-based expression of mammalian caveolin in Sf21 insect cells. A model system for the biochemical and morpho- logical study of caveolar biogenesis. *J. Biol. Chem.* **271,** 28,647–28,654.
39. Brada, D. and Dubach, U. C. (1984) Isolation of a homogeneous glucosidase II from pig kidney microsomes. *Eur. J. Biochem.* **141,** 149–156.
40. Vidugiriene, J., and Menon, A. K. (1993) Early lipid intermediates in GPI-anchor assembly are synthesized in the ER and located in the cytoplasmic leaflet of the ER membrane bilayer. *J. Cell Biol.* **121,** 987–996.

6

Analysis of Lipids in Caveolae

Jun Liu and Jan E. Schnitzer

1. Introduction

Caveolae are small invaginated microdomains located on the surface of many cells that form in part through the oligomerization of a rather unusual protein called caveolin ***(1,2)***. Caveolin appears to be embedded only partially in the lipid membrane with its N- and C-terminals exposed to the cytoplasm ***(3)***. Caveolin can form large homooligomeric spherical structures in solution and interacts with a variety of lipids and lipid-anchored proteins ***(4–6)***. Caveolae are rich in certain lipid-anchored proteins, such as nonreceptor tyrosine kinases and endothelial nitric oxide synthase ***(7,8)***, as well as lipids, such as, sphingomyelin, phosphoinositides and gangliosides, G_{M1} ***(7,9)***. Many of these same molecules have also been detected in various cell types in isolated Triton-insoluble membranes ***(10–13)***, which contain caveolae as well as various other membrane microdomains including those rich in GPI-anchored proteins or cytoskeletal elements ***(14)***. Interestingly, caveolae appear capable of discriminating between lipid-anchored proteins by excluding many GPI-anchored proteins under normal, nonperturbed conditions ***(14)***. However, after clustering induced by molecules that crosslink GPI-anchored proteins, they may enter caveolae to be internalized ***(15,16)***. The unusual lipid composition of caveolae may result not only from the presence of caveolin, which may act as a scaffolding protein for various signaling molecules ***(6)***, but also the inherent properties of these lipids that can self-associate and cluster into microdomains, thereby, creating a local environment that is distinct from the remaining membrane. Caveolae also have the molecular machinery to function as vesicular carriers mediating the endocytosis and transcytosis of select molecules into and across the endothelium ***(17,18)***. The selectivity of this traffick-

From: *Methods in Molecular Biology, Vol. 116: Protein Lipidation Protocols*
Edited by: M. H. Gelb © Humana Press Inc., Totowa, NJ

ing pathway provides a novel strategy for tissue- and disease-specific vascular targeting to overcome the endothelial cell barrier that normally prevents drug and gene delivery ***(19)***. In addition, caveolae may be important in multiple diseases including atherosclerosis by functioning in cholesterol hemostasis ***(20)*** and also acting as "mechanosensors" on the luminal surface of endothelial cells in vivo that may be quite consequential in the feedback regulation of blood flow ***(19,21)***.

Here we describe how caveolae are purified and used to determine their lipid compositions, with a primary focus on sphingomyelin and its metabolites. For the methods of extraction and analysis of inositol phospholipids, please *see* **refs. *7*** and ***22***. Although sphingomyelin has long been considered as a structural component of the plasma membrane, recent studies demonstrate that the metabolites of sphingomyelin play important roles in signal transduction. Cytokines, such as tumor necrosis factor-α and interleukin-1β ***(23,24)***, as well as ionizing radiation ***(25)*** activate sphingomyelinase which causes a rapid hydrolysis of sphingomyelin to generate ceramide acting as a second messenger. Ceramide, then, activates a membrane-bound protein kinase, named ceramide-activated protein kinase/kinase suppressor of Ras ***(26,27)***. Another direct target of ceramide is a cytosolic protein phosphatase ***(28)***. Our recent results have shown that purified caveolae are rich both in sphingomyelin as well as neutral sphingomyelinase ***(7,29)***, suggesting a potential role of caveolae in regulating sphingomyelin pathway.

2. Materials

2.1. Isolation of Caveolae for Labeling and Extraction of Sphingomyelin and Ceramide

1. Ringer's solution: 114 m*M* NaCl, 4.5 m*M* KCl, 1 m*M* $MgSO_4$, 11 m*M* Dextrose, Na_2HPO_4, 25 m*M* $NaHCO_3$.
2. MBS (pH 6.0): 20 m*M* 2-(N-Morpholino)ethanesulfonic acid, 125 m*M* NaCl.
3. 1% Colloidal silica particles (Vascular Genomics Inc., San Diego, CA) in MBS.
4. 0.1% Polyacrylic acid (Polyscience, Inc.) in MBS.
5. Buffer H: 250 m*M* sucrose, 25 m*M* Hepes, 2 m*M* KCl, pepstatin A (10 μg/mL), leupeptin (10 μg/mL), *o*-phenanthroline (10 μg/mL), 4-(2-aminoethyl) benzenesulfonyl fluoride (10 μg/mL), and *trans*-epoxysuccinyl-L-leucinamido(4-guanidino)butane (50 μg/mL). Protease inhibitors are added freshly from concentrated stock solutions (200X).
6. [^{3}H]Choline (81 Ci/mmol, NET-109, NEN/Du Pont).
7. 102% Nycodenz (w/v) (Accurate Chemical and Scientific) containing 20 m*M* KCl.
8. Type C and AA Teflon pestle/glass homogenizers (Thomas Scientific).
9. Bicinchoninic acid (BCA) protein assay kit (Pierce, Rockford, IL).
10. Lysis buffer: 1% Triton in MBS (pH 6.0) with 20 m*M* KCl, pepstatin A (10 μg/mL), leupeptin (10 μg/mL), *o*-phenanthroline (10 μg/mL), 4-(2-aminoethyl)

benzenesulfonyl fluoride (10 μg/mL), and *trans*-epoxysuccinyl-L-leucinamido (4-guanidino)butane (50 μg/mL). Protease inhibitors were added freshly from concentrated stock solutions (200X).

11. Solution E: chloroform:methanol:concentrated hydrochloric acid (100:100:1).
12. BSS: 135 m*M* NaCl, 4.5 m*M* KCl, 1.5 m*M* $CaCl_2$, 0.5 m*M* $MgCl_2$, 5.6 m*M* glucose, 10 m*M* HEPES, pH 7.2.
13. 100 m*M* EDTA.
14. Kreb's solution: 118.5 m*M* NaCl, 4.7 m*M* KCl, 1.18 m*M* $MgCl_2$, 1.18 m*M* KH_2PO_4, 25 m*M* $NaHCO_3$, 0.5 m*M* $CaCl_2$, 10 m*M* glucose.
15. Glass tubes: 12 × 75 mm; 13 × 100 mm.

2.2. Separation of Sphingomyelin by Thin-Layer Chromatography (TLC)

1. Solvent system: 50 mL chloroform, 30 mL methanol, 8 mL glacial acetic acid, 3 mL deionized water.
2. Chloroform:methanol (1:1) (v/v).
3. Sphingomyelin (from bovine brain, Sigma).
4. Phosphatidylcholine (from bovine brain, Sigma).
5. 0.1 *M* Potassium chloride in methanol.
6. Silica gel 60 TLC plates: 20 × 20 cm, 0.25-mm thickness without fluorescent indicator (Merck, obtained from Fisher).
7. Iodine.
8. Hamilton syringe for application of samples.
9. Hair dryer.
10. Filter paper-lined chromatography tank with lid.

2.3. Phosphorylation of Ceramide by Diacylglycerol (DAG) Kinase

1. Imidazole (Sigma).
2. Diethylenteriaminepentaacetic acid (DETAPAC) (Sigma).
3. Octyl-β-D-glucopyranoside (Sigma).
4. Cardiolipin (Avanti Polar Lipids).
5. *sn*-1,2-Diacylglycerol kinase (from *Escherichia coli*, cat. no. 266724, Calbiochem).
6. Eramide (Sigma): 5 mg/mL in chloroform:methanol (1:1).
7. [γ-^{32}P] Adenosine triphosphate (ATP) (3000 Ci/mmol, NEN/DuPont).
8. Solution E: chloroform:methanol:concentrated hydrochloric acid (100:100:1).
9. BSS (*see* **Subheading 2.1,. item 12**).
10. 100 m*M* EDTA.
11. Glass tubes: 12 × 75mm, 13 × 100mm.
12. BCA assay kit (Pierce, Rockford, IL).

2.4. Separation of Ceramide-1-Phosphate (Cer-1-P) by TLC

1. Solvent system: 65 mL chloroform, 15 mL methanol, 5 mL acetic acid (Glacial).
2. Chloroform:methanol (1:1) (v/v).

3. Ceramide (Sigma).
4. Silica gel 60 TLC plates: 20 × 20 cm, 0.25-mm thickness without fluorescent indicator (Merck, obtained from Fisher).
5. Iodine.
6. Hamilton syringe for application of samples.
7. Hair dryer.
8. Filter paper-lined chromatography tank with lid.
9. Kodak X-ray films.

3. Methods

3.1. Isolation of Caveolae for Labeling and Extraction of Sphingomyelin and Ceramide

Both sphingomyelin and ceramide are minor membrane sphingolipids and are barely detectable from plasma membranes and their caveolae by conventional methods. Hence, we use radiolabels to accurately quantify the levels of the lipids in endothelial cell plasma membranes and their caveolae.

Three extraction procedures, i.e., the Bligh and Dyer extraction ***(30)***, the perchloric acid (PCA), and the trichloroacetic acid (TCA) procedures, may be used for recovery of radiolabeled products. Among these, the Bligh and Dyer extraction procedure is the most commonly used method for maximum recovery of the radiolabeled lipids. If the extraction is to be conducted on plastic culture plates, first use methanol to remove cell and lipids from the plates into a glass tube and then mix with chloroform to extract lipids. All samples should be processed immediately, and kept on ice during the extraction procedures. When dried down, lipid extracts are reasonably stable at –20°C and can be stored frozen for several days.

The following procedures are described for radiolabeling and extraction of lipids from subcellular fractions including caveolae of rat lung vascular endothelium *in situ* as well as endothelial cell monolayers grown in culture. To obtain sufficient signals from other types of cells or tissues, radiolabeling conditions can be optimized by changing radioisotope concentration and/or the period of time for labeling.

3.1.1. Labeling of Rat Lung with [^{3}H]Choline for Caveolae Isolation and Lipid Extraction

Recently, we have developed a silica-coating technique for purifying endothelial cell plasma membranes and their caveolae to homogeneity directly from rat lung tissue ***(14,31)***. This procedure is described here briefly. We also described the methods of radiolabeling rat lung tissue and isolating and extracting the lipids from the purified endothelial cell plasma membrane and their caveolae.

1. Perfuse rat lung microvasculature via the pulmonary artery with 20 mL of Ringer's solution at 4°C at 6–10 mmHg to remove the blood followed by MBS and then 8 mL of 1% colloidal silica solution. After flushing with MBS, the silica particles coating the luminal cell surface of endothelium are crosslinked by perfusing 6 mL of polyacrylic acid solution. After crosslinking, the rat lungs are perfused with 8 mL Ringer's solution.
2. Excise the lungs, remove large blood vessels and bronchi, and chop the lung into $0.5 \times 0.5 \times 0.5$ cm^3 pieces.
3. Incubate the tissue sections in 25 mL Kreb's solution containing 200 µCi [^{3}H]choline with 5% CO_2/O_2 at 37°C for 3 h.
4. Wash the lung tissue three times with 5 mL ice-cold phosphate-buffered saline (PBS) to remove free [^{3}H]choline.
5. Homogenize lung tissue in 20 mL buffer H at 4°C for 12 strokes with a Dounce homogenizer (clearance, 15–23 mm). After filtering through a 0.53 µm Nytex net followed by a 0.3 µm, mix the homogenate with 102% Nycodenz to make a 50% final solution and load carefully the homogenate mixture over a 55–70% Nycodenz gradient. After centrifugation at 15,000 rpm (SW28 rotor, Beckman), 4°C for 30 min, collect the pellet which contains the silica-coated endothelial cell plasma membranes.
6. Resuspend the pellet in 1 mL MBS with or without 1% Triton X-100. Homogenize the silica-coated plasma membrane at 4°C for 20 (with Triton) or 50 (without Triton) strokes with a AA Teflon pestle/glass homogenizer (clearance, 0.10–0.15 mm) to shear off caveolae from the plasma membranes. Mix the homogenate with 60% sucrose containing 20 m*M* KCl to make a 40% final solution and layer on top of the sample with 35–0% sucrose gradient. After centrifugation at 30,000 rpm (SW55 rotor, Beckman), 4°C overnight, collect caveolae fraction which is a clear membrane band located around 15% sucrose of the gradient. Mix the fraction with 2 vol of MBS and spin down caveolae at a full speed for 2 h at 4°C with microcentrifuge.
7. Run protein assay using BCA assay kit to determine protein concentration in the plasma membranes and caveolae.
8. To extract lipids, place 50 µg plasma membrane or 10 µg caveolae into 12×75 mm glass tubes. Add 270 µL BSS, 30 µL 100 m*M* EDTA, and 1 mL solution E.
9. Vortex for 30 s and spin at 2000 rpm for 5 min to separate organic phase from aqueous. Transfer 0.4 mL organic (lower) phase to 12×75 mm tube (the total volume of lower phase is about 0.55 mL). Dry down under N_2, which may take about 10–12 min. Lipids can be stored at this point at –20°C for several days if desired.

3.1.2. Labeling of Endothelial Cells with [^{3}H]Choline and Lipid Extraction (Cell Monolayer)

Preincubation of cells with [^{3}H]palmitate labels not only sphingomyelin and ceramide, but many other sphingolipids and glycerolipids. An alternative approach is to prelabel cells with [^{3}H]choline. Both sphingomyelin and phosphatidylcholine become labeled in the choline moiety. These radiolabeling procedures require several days in order to achieve the labeling of sphingomyelin and ceramide to radioisotopic equilibrium. Only under these conditions can changes in radioactivity be interpreted as changes in mass.

1. Trypsinize confluent bovine aortic endothelial cells (BAEC) and resuspend the cells in a T-75 cm^2 flask or 6-well plate at the density of 0.5×10^6 cell/mL in DMEM (cat. no. 10316-024, Gibco-BRL) containing 10% fetal calf serum (FCS). To achieve a maximum signal, start to radiolabel the cells at a subconfluent stage by adding [^{3}H]choline to the incubation medium to a final concentration of 1 µCi/mL and continue to label for 3 d.
2. Remove the radiolabeling medium and wash the cells with PBS to remove free [^{3}H]choline.
3. Place the flask on ice and incubate the cells with 2.0 mL lysis buffer for 15 min. Collect the cell lysate and wash the flask with 2.0 mL lysis buffer. Combine the wash with the cell lysate.
4. Isolate the Triton-insoluble membrane fraction (TIM) as described originally ***(11)*** and in our past work ***(14)***, which is rich in caveolin but also contains other noncaveolar microdomains ***(14)***.
5. To extract lipid, place 50 µg TIM into 12 × 75 mm glass tube and add 1 mL solution E, 270 µL BSS and 30 µL 100 m*M* EDTA.
6. When measuring sphingomyelin in whole cell, place the 6-well plate on ice, wash the cells as in **step 2**. Add 2 mL methanol to each well and leave on ice for 10 min. Scrape the cells from the plate and transfer the resulting mix to 13 × 100 mm glass tubes. Wash the plate with 1.1 mL BSS, 120 µL 100 m*M* EDTA, and combine this mixture with the methanol plus cells. To extract the lipids, add 2 mL chloroform to the tubes.
7. Vortex for 30 s, spin at 2000 rpm for 5 min.
8. Transfer 1.8 mL lower (organic) phase to 12 × 75 mm tube (the total volume of lower phase is about 2.1 mL). Dry down under N_2. Lipids can be stored at this point at –20°C for several days if desired.

3.2. Separation of Sphongomyelin by TLC

Both sphingomyelin and phosphatidylcholine are radiolabeled after incubation of cells with [^{3}H]choline. Hence, it is necessary to further separate and identify sphingomyelin from other radiolabeled phospholipids. TLC is a commonly used method for lipid separation. The procedures described below give a clear separation of sphingomyelin from other lipids in one dimension.

3.2.1. Alkaline Hydrolysis

For the detection of sphingolipids including sphingomyelin, the lipid extract is subjected to alkaline hydrolysis to remove glycerophospholipids ***(32)***.

1. Add 500 µL 0.1 *M* potassium chloride in methanol to the tube containing lipid film (*see* **Subheadings 3.1.1.** and **3.1.2.**) and cover the tube with parafilm. Incubate at 37°C for 1 h with shaking (do not incubate for more than 2 h).

2. Add to the tube in the following order: 500 μL chloroform, 270 μL BSS, 30 μL 100 m*M* EDTA.
3. Vortex for 30 s and spin at 2000 rpm for 5 min. Transfer 400 μL lower phase to a new glass tube. Dry down under N_2. Separate the lipids as described in the following section.

3.2.2. Separation of Lipids by TLC

1. Add solvent system to the chromatography tank lined with filter paper to give a depth of 0.5–1.0 cm. Place a lid on the tank to allow equilibrium.
2. Draw a line with a pencil across the plate 1.5 cm from the bottom edge of the plate. Mark crosses on this line at 2-cm intervals.
3. Dilute sphingomyelin standard in chloroform:methanol (1:1) at the concentration of 0.5 mg/mL.
4. Resuspend lipid samples in 50 μL chloroform:methanol (1:1) containing 25 μg sphingomyelin. Apply the sample onto a cross point of the TLC plate by using a Hamilton syringe. Rinse the tube with 20 μL chloroform:methanol and apply to the same spot.
5. Repeat **step 4** for the other samples and the standard, sphingomyelin. Rinse the Hamilton syringe between each sample. Dry the plate using a hair dryer.
6. Separate the lipids by placing the plate in the solvent. Let it run until solvent front is 1–2 cm from top. Remove plate and allow to air-dry in a fume hood.
7. When the plate is free of all traces of solvent, place the plate in the tank containing sublimated iodine until the sphingomyelin spots are visible. Mark the positions of these spots.
8. To measure radioactivity in each spot, scrape the marked areas and remove to scintillation vials. Add 4 mL scintillation solution to each vial, mix and count in a scintillation counter.

3.3. Phosphorylation of Ceramide by DAG Kinase

The classical approaches to analyze ceramides are based on their isolation by TLC, followed by hydrolysis and analysis of the liberated and derivative sphingoid bases or fatty acids by gas chromatography (GC) ***(33)*** or high-performance liquid chromatography (HPLC) ***(34)***. Since ceramides are minor membrane sphingolipids, a large quantity of membranes is required to determine the level of ceramide by using these methods. An alternative approach is to phosphorylate in vitro ceramide to become Cer-1-P by DAG kinase followed by TLC separation of the phosphorylated product ***(35)***. This simple and sensitive assay can detect the level of ceramide as low as 50 pmoles.

3.3.1. Extraction of Ceramide from Tissues or Cell in Culture

1. Add 100 μg silica-coated endothelial cell plasma membrane isolated from rat lung (*see* **Subheading 3.1.1.**) to 12 × 75 mm glass tubes, spin at 2000 rpm for 10 min. To the pellet, add 270 μL BSS, 30 μL 100 m*M* EDTA, and 1 mL solution E.

2. Vortex for 30 s and spin at 2000 rpm for 5 min to separate organic phase from aqueous. Transfer 0.4 mL organic (lower) phase to 12 × 75 mm tubes (the total volume of lower phase is about 0.55 mL). Dry down under N_2, which takes about 10–12 min. Lipids can be stored at this point at –20°C for several days if desired.
3. When measuring ceramide in cultured cells, place the 6-well plate on ice, wash the cells with PBS. Add 2 mL methanol to each well and leave on ice for 10 min. Scrape the cells from the plate and transfer the resulting mix to 13 × 100 mm glass tubes. Wash the plate with 1.1 mL BSS, 120 μL 100 m*M* EDTA, and combine this mixture with the methanol plus cells. To extract the lipids, add 2 mL chloroform to the tubes.
4. Vortex for 30 s and spin at 2000 rpm for 5 min.
5. Transfer 400 μL lower (organic) phase to 12 × 75 mm tubes (for measuring ceramide in cultured cells, transfer 1.8 mL lower phase to 12 × 75 mm tubes). Dry down under N_2. Lipids can be stored at this point at –20°C for several days if desired.

3.3.2. Alkaline Hydrolysis

1. Add 500 μL 0.1 *M* potassium chloride in methanol to lipid film, cover with parafilm. Incubate at 37°C for 1 h with shaking (do not incubate for longer than two hours). Ceramide standards (*see* **Note 1**) should be also included from this step onwards.
2. Add to the tube in the following order: 500 μL chloroform, 270 μL BSS, 30 μL 100 m*M* EDTA.
3. Vortex for 30 s and spin at 2000 rpm for 5 min. Transfer 400 μL lower phase to new glass tubes. Dry down under N_2. Phosphorylate the lipids by DAG kinase as described in the following section.

3.3.3. Phosphorylation of Ceramide by DAG kinase

1. To the lipid film, add 100 μL reaction buffer containing 50 m*M* imidazole, pH 6.5, 50 m*M* NaCl, 12.5 m*M* $MgCl_2$, 1 m*M* EDTA, 5 m*M* cardiolipin, 1.5% octyl-β-D-glucopyranoside, 0.2 m*M* DETAPAC, 2 m*M* dithiothreitol (DTT), 4 mg DAG kinase, and 1 m*M* adenosine triphosphate (ATP).
2. Initiate phospho-radiolabeling by adding 1 μL [γ-^{32}P]ATP (3000 Ci/mmol) into each tube.
3. After incubation at 25°C for 30 min, stop the reaction by adding 1 mL solution E, 170 μL BSS, and 30 μL 100 m*M* EDTA.
4. Vortex for 30 s and spin at 2000 rpm for 5 min. Transfer 400 μL lower phase to new glass tube. Dry down under N_2. Separate the radiolabeled lipids as described in the following section.

3.4. Separation of Cer-1-P by TLC

Incubation of DAG kinase with membrane lipid extract results in phosphorylation of both ceramides and DAG. Hence, it is necessary to further separate and identify Cer-1-P from other phosphorylated lipids. TLC is a commonly used method for lipid separation. The procedures described below give a clear separation of Cer-1-P from other lipids in one dimension.

3.4.1. Resolution of Cer-1-P

1. Add the solvent system to the chromatography tank lined with filter paper to give a depth of 0.5–1.0 cm. Place a lid on the tank to allow equilibrium.
2. Draw a line with a pencil across the plate 1.5 cm from the bottom edge of the plate. Mark crosses on this line at 2-cm intervals.
3. Resuspend lipid sample in 50 µL chloroform:methanol (1:1). Apply the sample onto a cross point of the TLC plate by using a Hamilton syringe. Rinse the tube with 20 µL chloroform:methanol and apply to the same spot.
4. Repeat **step 3** for the other samples and the standard, ceramide. Rinse the Hamilton syringe between each sample. Dry the plate using a hair dryer.
5. Separate the lipids by placing the plate in the solvent. Let it run until solvent front is 1–2 cm from top. Remove plate and allow to air-dry in a fume hood.
6. When the plate is free of all traces of solvent, place the plate in the tank containing sublimated iodine until the ceramide spots are visible. Mark the positions of these spots.
7. Expose plate to X-ray film (individually rapped) from 1 h to overnight.
8. Cut the radioactive bands on the plate corresponding to Cer-1-P and count in a scintillation counter. Convert cpm into picomoles by reading values off a simultaneously run standard curve consisting of known quantities of ceramide.

4. Note

Ceramide standards are made as follows: Dissolve ceramide in chloroform:methanol (1:1) to yield a stock solution of 10^{-2} *M* ceramide. Dilute this stock serially to generate 10^{-5} and 10^{-6} *M* ceramide solution.

Ceramide (pmol)	Amount needed
50	50 µL of 10^{-6} *M* ceramide
100	100 µL of 10^{-6} *M* ceramide
200	20 µL of 10^{-5} *M* ceramide
400	40 µL of 10^{-5} *M* ceramide
800	80 µL of 10^{-5} *M* ceramide
1600	160 µL of 10^{-5} *M* ceramide

Acknowledgment

We thank Dr. Richard Kolesnick for his helpful discussion about this manuscript.

References

1. Rothberg, K. G., Heuser, J. E., Donzell, W. C., Ying, Y. S., Glenney, J. R., and Anderson, R. G. (1992) Caveolin, a protein component of caveolae membrane coats. *Cell* **68**, 673–682.

2. Sargiacomo, M., Scherer, P. E., Tang, Z., Kubler, E., Song, K. S., Sanders, M. C., and Lisanti, M. P. (1995) Oligomeric structure of caveolin: implications for caveolae membrane organization. *Proc. Natl. Acad. Sci. USA* **92**, 9407–9411.
3. Monier, S., Parton, R. G., Vogel, F., Henske, A., and Kurzchalia, T. V. (1995) VIP21-caveolin, a membrane protein constituent of the caveolar coat, oligomerizes in vivo and in vitro. *Mol. Biol. Cell* **6**, 911–927.
4. Murata, M., Kurzchalia, T., Peranen, J., Schreiner, R., Wieland, F. T., and Simons, K. (1995) VIP21-caveolin is a cholesterol-binding protein. *Proc. Natl. Acad. Sci. USA* **92**, 10,339–10,343.
5. Fra, A. M., Masserini, M., Palestini, P., Sonnino, S., and Simons, K. (1995) A photo-reactive derivative of ganglioside GM1 specifically cross-links VIP21-caveolin on the cell surface. *FEBS Lett* **375**, 11–14.
6. Li, S., Couet, J., and Lisanti, M. P. (1996) Src tyrosine kinases, Galpha subunits, and H-Ras share a common membrane-anchored scaffolding protein, caveolin. Caveolin binding negatively regulates the auto-activation of Src tyrosine kinases. *J. Biol. Chem.* **271**, 29,182–29,190.
7. Liu, J., Oh, P., Horner, T., Rogers, R. A., and Schnitzer, J. E. (1997) Organized endothelial cell surface signal transduction in caveolae distinct from glycosylphosphatidylinositol-anchored protein microdomains. *J. Biol. Chem.* **272**, 7211–7222.
8. Rizzo, V., McIntosh, D. P., Oh, P., and Schnitzer, J.E . (1998) In situ flow activation of luminal endothelial cell surface eNOS in caveolae (Submitted).
9. Parton, R. G. (1994) Ultrastructural localization of gangliosides: GM1 is concentrated in caveolae. *J. Histochem. Cytochem.* **42**, 155–166.
10. Pike, L. J. and Casey, L. (1996) Localization and turnover of phosphatidylinositol 4,5-bisphosphate in caveolin-enriched membrane domains. *J. Biol. Chem.* **271**, 26,453–26,456.
11. Brown, D. A. and Rose, J. K. (1992) Sorting of GPI-anchored proteins to glycolipid-enriched membrane subdomains during transport to the apical cell surface. *Cell* **68**, 533–544.
12. Liu, P. and Anderson, R. G. (1996) Compartmentalized production of ceramide at the cell surface. *J. Biol. Chem.* **270**, 27,179–27,185.
13. Sargiacomo, M., Sudol, M., Tang, Z., and Lisanti, M. P. (1993) Signal transducing molecules and glycosyl-phosphatidylinositol-linked proteins form a caveolin-rich insoluble complex in MDCK cells. *J. Cell Biol.* **122**, 789–807.
14. Schnitzer, J. E., McIntosh, D. P., Dvorak, A. M., Liu, J., and Oh, P. (1995) Separation of caveolae from associated microdomains of GPI-anchored proteins. *Science* **269**, 1435–1439.
15. Mayor, S., Rothberg, K. G., and Maxfield, F. R. (1994) Sequestration of GPI-anchored proteins in caveolae triggered by cross-linking. *Science* **264**, 1948–1951.
16. Horner, T. and Schnitzer, J. E. (1996) Differential endocytosis of antibodies to caveolae-specific proteins vs. GPI-anchored proteins. *Mol. Biol. Cell* **7(suppl)**, 265a.

17. Schnitzer, J. E., Liu, J., and Oh, P. (1995) Endothelial caveolae have the molecular transport machinery for vesicle budding, docking, and fusion including VAMP, NSF, SNAP, annexins, and GTPases. *J. Biol. Chem.* **270,** 14,399–14,404.
18. Schnitzer, J. E., Oh, P., Pinney, E., and Allard, J. (1994) Filipin-sensitive caveolae-mediated transport in endothelium: reduced transcytosis, scavenger endocytosis, and capillary permeability of select macromolecules. *J. Cell Biol.* **127,** 1217–1232.
19. Schnitzer, J. E. (1997) The endothelial cell surface and caveolae in health and disease, in: *Vascular Endothelium: Physiology, Pathology and Therapeutic Opportunities*. (Born, G. V. R. and Schwartz, C. J., eds.), Schattauer, Stuttgart, pp. 77–95.
20. Fielding, P. E. and Fielding, C. J. (1995) Plasma membrane caveolae mediate the efflux of cellular free cholesterol. *Biochemistry* 34, 14,288–14,292.
21. Rizzo, V., Sung, A., Oh, P., and Schnitzer, J. E. (1998) Rapid mechanotransduction in situ at the luminal cell surface of vascular endothelium and its caveolae. *J. Biol. Chem.* (in press).
22. Bird, I. M. (1994) Analysis of cellular phosphoinositides and phosphoinositols by extraction and simple analytical procedures, in *Methods in Molecular Biology*, vol. 27, (Graham, J. M. and Higgins, J. A., eds.), Humana Press, Totowa, NJ, pp. 227–248.
23. Kolesnick, R., and Golde, D. W. (1994) The sphingomyelin pathway in tumor necrosis factor and interleukin-1 signaling. *Cell* **77**, 325–328.
24. Hannun, Y. A. (1996) Function of ceramide in coordinating cellular responses to stress. *Science* **274**, 1855–1859.
25. Haimovitz-Friedman, A., Kan, C. C., Ehleiter, D., Persaud, R. S., McLoughlin, M., Fuks, Z., and Kolesnick, R. N. (1994) Ionizing radiation acts on cellular membranes to generate ceramide and initiate apoptosis. *J. Exp. Med.* **180**, 525–535.
26. Liu, J., Mathias, S., Yang, Z., and Kolesnick, R. N. (1994) Renaturation and tumor necrosis factor-alpha stimulation of a 97-kDa ceramide-activated protein kinase. *J. Biol. Chem.* **269**, 3047–3052.
27. Zhang Y, Yao, B, Delikat, S, Bayoumy, S., Lin, X.H., Basu, S., McGinley, M., Chan-Hui, P. Y., Lichenstein, H., and Kolesnick, R. (1997) Kinase suppressor of Ras is ceramide-activated protein kinase. *Cell* **89**, 63–72.
28. Dobrowsky, R. T. and Hannun, Y. A. (1992) Ceramide stimulates a cytosolic protein phosphatase. *J. Biol. Chem.* **267**, 5048.
29. Liu, J. and Schnitzer, J. E. (1998) Flow-dependent activation of neutral but not acidic sphingomyelinase in endothelial cell caveolae (manuscript in preparation).
30. Bligh, E. G. and Dyer, W. J. (1959) A rapid method for total lipid extraction and purification. *Can. J. Biochem. Physiol.* **37**, 911–917.
31. Schnitzer, J. E., Oh, P., Jacobson, B. S., and Dvorak, A. M. (1995) Caveolae from luminal plasmalemma of rat lung endothelium: microdomains enriched in caveolin, Ca^{2+}-ATPase, and inositol trisphosphate receptor. *Proc. Natl. Acad. Sci. USA* **92**, 1759–1763.
32. Dressler, K. A., Mathias, S., and Kolesnick, R. N. (1992) Tumor necrosis factor-alpha activates the sphingomyelin signal transduction pathway in a cell-free system. *Science* **255**, 1715–1718.

33. Samuelson, B. and Samuelson, K. (1969) *J. Lipid Res.* **10**, 41–46.
34. Jungalwala, F. B., and Evans, J. E., Bremer, E., and McCluer, R. H. (1983) Analysis of sphingoid bases by reversed-phase high performance liquid chromatography. *J. Lipid Res.* **24**, 1380–1388.
35. Preiss, J., Loomis, C. R., Bishop, W. R., Stein, R., Niedel, J. E., and Bell, R. M. (1986) Quantitative measurement of sn-1,2-diacylglycerols present in platelets, hepatocytes, and ras- and sis-transformed normal rat kidney cells. *J. Biol. Chem.* **261**, 8597–8600.

7

Analysis of the Carbohydrate Components of Glycosylphosphatidylinositol Structures Using Fluorescent Labeling

Nicole Zitzmann and Michael A. J. Ferguson

1. Introduction

Glycosylphosphatidylinositols (GPIs) are a family of structures that contain the structural motif: Manα1-4GlcNH$_2\alpha$1-6*myo*-Inositol-1-PO$_4$-lipid. This common substructure suggests that this family of molecules is biosynthetically related, and it differentiates them from other glycosylated phosphoinositides, such as the glycosylated phosphatidylinositols of mycobacteria and the glycosylated inositol phosphoceramides of yeasts and plants. The GPI family can be conveniently divided into two groups, based on structural homology and function. The first group *(1–34)* is the membrane protein anchors (**Fig. 1**) which are found covalently linked to the C-termini of a wide variety of externally disposed plasma-membrane proteins throughout the eukaryotes. These GPI anchors afford a stable attachment of proteins to the membrane, and can be viewed as an alternative mechanism of membrane attachment to a single-pass hydrophobic transmembrane peptide domain. For recent reviews of GPI anchor structure, biosynthesis, and function, *see* **refs. *35–40***. The second group of GPI structures have only been found in protozoan organisms. These molecules exist as free glycophospholipids, such as the glycoinositol phospholipids (GIPLs) of the *Leishmania, Trypanosoma cruzi, Leptomonas, Herpetomonas, Phytomonas,* and *Toxoplasma* ***(35,41–45)***, or are attached to phosphorylated repeating units, as in the lipophosphoglycans (LPGs) of the *Leishmania* ***(35,46)***. This chapter describes protocols specifically designed to analyze the carbohydrate components of protein-linked GPI anchors, although they are also applicable to the GIPLs and, to some extent, the LPGs.

From: *Methods in Molecular Biology, Vol. 116: Protein Lipidation Protocols*
Edited by: M. H. Gelb © Humana Press Inc., Totowa, NJ

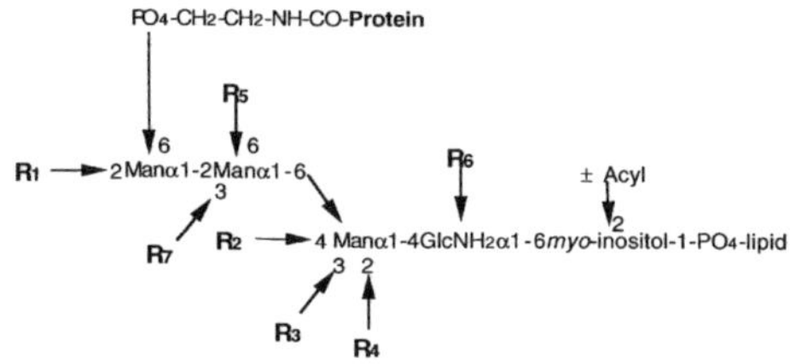

Protein	R_1	R_2	R_3	R_4	R_5	R_6	R_7	Lipid	Inositol-acyl	
T.brucei VSG	±αGal		αGal_{2-4}				±βGal	diacylglycerol		*(1)*
T.congolense VSG		Galβ1-6βGlcNAc						diacylglycerol		*(2)*
T.brucei PARP		[sialylated polylactosamine sidechain]						*lyso*-acylglycerol	+	*(3, 4)*
H. davidi 21, 31, 45 kDa protein			n.d.					*lyso*-alkylglycerol		*(5)*
T. cruzi 1G7 antigen	αMan							alkylacylglycerol and ceramide		*(6, 7)*
T. cruzi Ssp-4 antigen			n.d.					ceramide		*(8)*
T.cruzi Tc85								alkylacylglycerol	palmitoyl	*(9, 10)*
T.cruzi mucins	αMan	[	± Hex	]		AEP		alkylacylglycerol and ceramide		*(11, 12)*
Leishmania PSP								alkylacylglycerol		*(13, 14)*
Toxoplasma gp23	βGalNAc							diacylglycerol		*(15)*
Plasmodium MSA-1	αMan							diacylglycerol	± myristoyl	*(16)*
Paramecium VSA		[	Manα1-PO_4-	]				ceramide		*(17)*
S.cerevisiae anchors	αMan_{1-2}							ceramide and diacylglycerol		*(18)*
D.discoideum PsA	±αMan	[1-2 $EtNPO_4$, positions unknown]						ceramide		*(19)*
Torpedo AChE	αGlc	±βGalNAc		$EtNPO_4$	±$EtNPO_4$?			diacylglycerol		*(20, 21)*
Rat Thy-1	±αMan	βGalNAc		$EtNPO_4$				n.d.		*(22)*
Hamster Prp	±αMan	[±NANA-Hex-HexNAc / $EtNPO_4$]						n.d.		*(23)*
Mouse NCAM	±αMan	±βGalNAc		n.d.	n.d.			n.d.		*(24)*
Bovine 5'-NT	±αMan	±HexNAc?		$EtNPO_4$?	±$EtNPO_4$?			n.d.		*(25)*
Bovine AChE		[HexNAc, $EtNPO_4$, position unknown]						alkylacylglycerol		*(26)*
Porcine MDP		±Galβ1-3GalNAc		$EtNPO_4$	±$EtNPO_4$?			diacylglycerol		(27)
Human MDP	±αMan	±Galβ1-3GalNAc		n.d.	n.d.			n.d.		*(27)*
Human AChE				$EtNPO_4$?	±$EtNPO_4$			alkylacylglycerol	palmitoyl	*(28)*
Human PLAP		[		$EtNPO_4$	]			alkylacylglycerol		*(29)*
Human CD52	±αMan			$EtNPO_4$?	±$EtNPO_4$?			diacylglycerol	±palmitoyl	(30)
Human CD59	±αMan	±βGalNAc		$EtNPO_4$?	±$EtNPO_4$?			alkylacylglycerol	palmitoyl	*(31-33)*

Fig. 1. GPI anchor structures. All GPI anchors attached to protein contain the conserved structure shown above, with various substituents (R_1–R_7) and lipids, as indicated. Some structures contain an additional fatty acyl chain attached to the 2-position of the *myo*-inositol ring. All metazoan organisms contain at least one, and sometimes two, extra ethanolamine phosphate ($EtNPO_4$) substituents, in addition to the one used as a bridge to the protein C-terminal amino acid. When a substituent is known to be attached to a certain sugar residue, but the linkage position is unknown, this is indicated by a question mark. Square brackets are used to show substituents for which the site of attachment has not been determined. The ± symbol indicates that the associated residue is found on only a proportion of the structures. AEP is 2-aminoethylphosphonate.

Electrospray mass spectrometric methods have previously been described to characterize the phosphatidylinositol components of GPI anchors ***(47)*** and chromatographic methods to characterize the glycan components of GPI anchors ***(47–49)***. However, the previous methods for glycan characterization relied on introducing a tritium radiolabel into the GPI glycan, with all of the associated complications of handling NaB^3H_4 reductant. Here is described a new method for labeling GPI anchor glycans using the fluorophore 2-aminobenzamide (2-AB), which can be used both on purified glycoproteins and on small (microgram) quantities of glycoproteins immobilized on polyvinylidene difluoride (PVDF) membranes after Western blotting.

The 2-AB labeling procedure described depends on the presence of non-N-acetylated glucosamine (GlcN) in all GPI structures. The free amino group of the GlcN residue can be exploited by the nitrous acid deamination reaction, which converts the GlcN to 2,5-anhydromannose (AHM) and simultaneously releases the inositol residue (**Fig. 2**). The aldehyde group of the AHM residue is then coupled to 2-AB by reductive amination, to give a stable covalent linkage between the GPI glycan and the fluorophore. The protein and ethanolamine substituents are subsequently removed by aqueous HF dephosphorylation to yield 2-AB-labeled GPI neutral glycans that can be characterized by liquid chromatography. Examples of the results obtained using a pure glycoprotein standard, a variant surface glycoprotein (VSG) of *Trypanosoma brucei* ***(1)***, are shown in **Fig. 3** and the results obtained using a Western blot band of this same glycoprotein are shown in **Fig. 4**.

2. Materials

1. High-pressure liquid chromagraphy (HPLC)-grade water.
2. Screw-top Eppendorf tubes (e.g., BDH-Merck, Poole, UK).
3. 0.5-mL Microtubes (BDH-Merck).
4. 48% Aqueous hydrofluoric acid (HF), Aristar-grade (BDH-Merck). Store in 0.5-mL aliquots in Eppendorf tubes at –20°C. **Caution:** Highly corrosive.
5. Dewar container (e.g., BDH-Merck).
6. Access to a freeze-drying apparatus (e.g., BDH-Merck).
7. Access to a sonicating water bath (e.g., BDH-Merck).
8. 0.3 *M* Sodium acetate buffer, pH 4.0. Made by titrating 0.3 *M* sodium acetate solution to pH 4.0 with glacial acetic acid. Stable at room temperature for several months.
9. 1 *M* Sodium nitrite. Always prepare fresh just before use.
10. C8 and NH_2 Isolute™ cartridges (IST, Mid-Glamorgan, UK).
11. Methanol, HPLC-grade (BDH-Merck).
12. Dowex AG50X12, 200–400 mesh (Bio-Rad, Hemel Hempstead, UK), converted to the H^+ form by washing with >10 vol 1 *M* HCl and >20 vol water. Store with an equal volume of water at 4°C.
13. Dowex AG3X4, 200–400 mesh (Bio-Rad), converted to the OH^- form by washing with >10 vol 1 *M* NaOH and >20 vol water. Store with an equal volume of water at 4°C.
14. Access to a Speed-Vac (Savant Instruments, Holbrook, NY) or rotary evaporator (Jencons, Leighton Buzzard, UK).

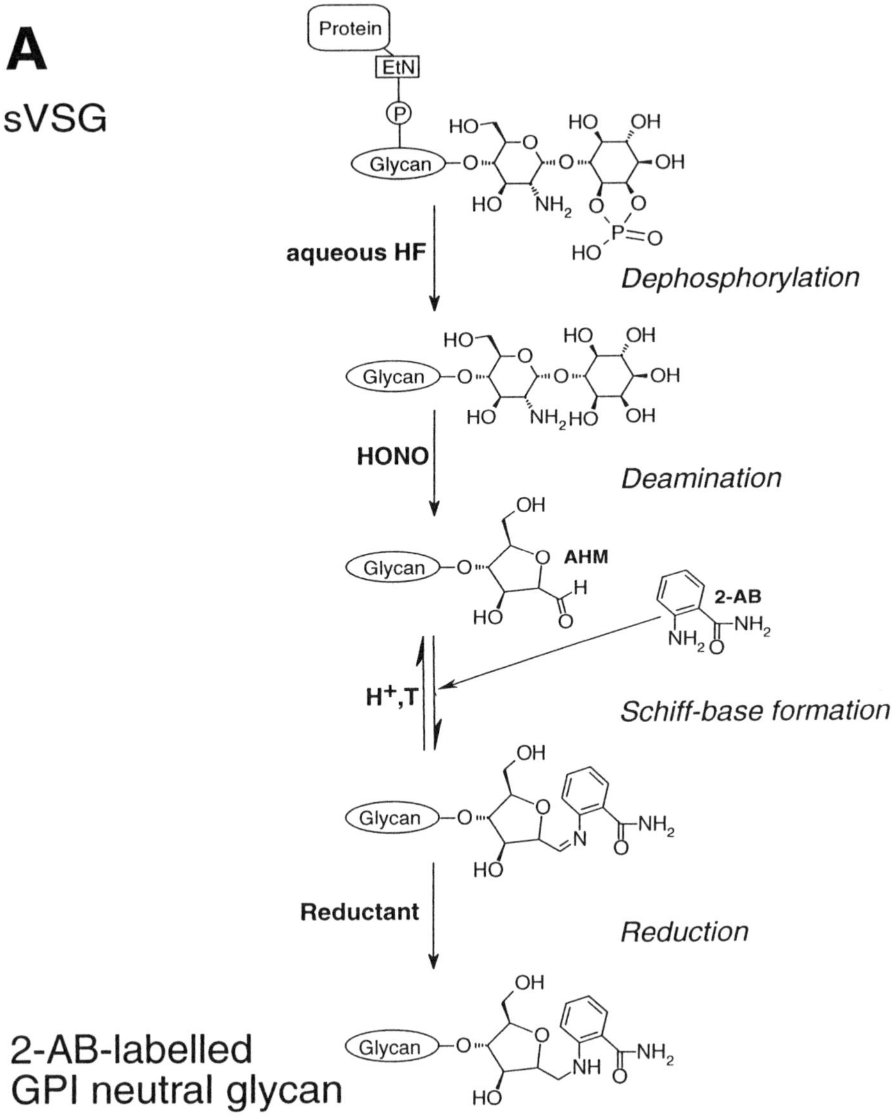

Fig. 2. General schemes for preparing the 2-AB-labeled GPI neutral glycans. (**A**) The reaction scheme for producing 2-AB-labeled GPI neutral glycans from sVSG, or any other GPI-anchored glycoprotein, using purified glycoprotein in solution ; the in-the-tube method *(continued)*

15. Signal™ 2-AB labeling kit (Oxford GlycoSystems, Abingdon, UK).
16. Access to a heating block (e.g., BDH-Merck).
17. 3MM Whatman paper (Whatman International Ltd., Maidstone, UK).

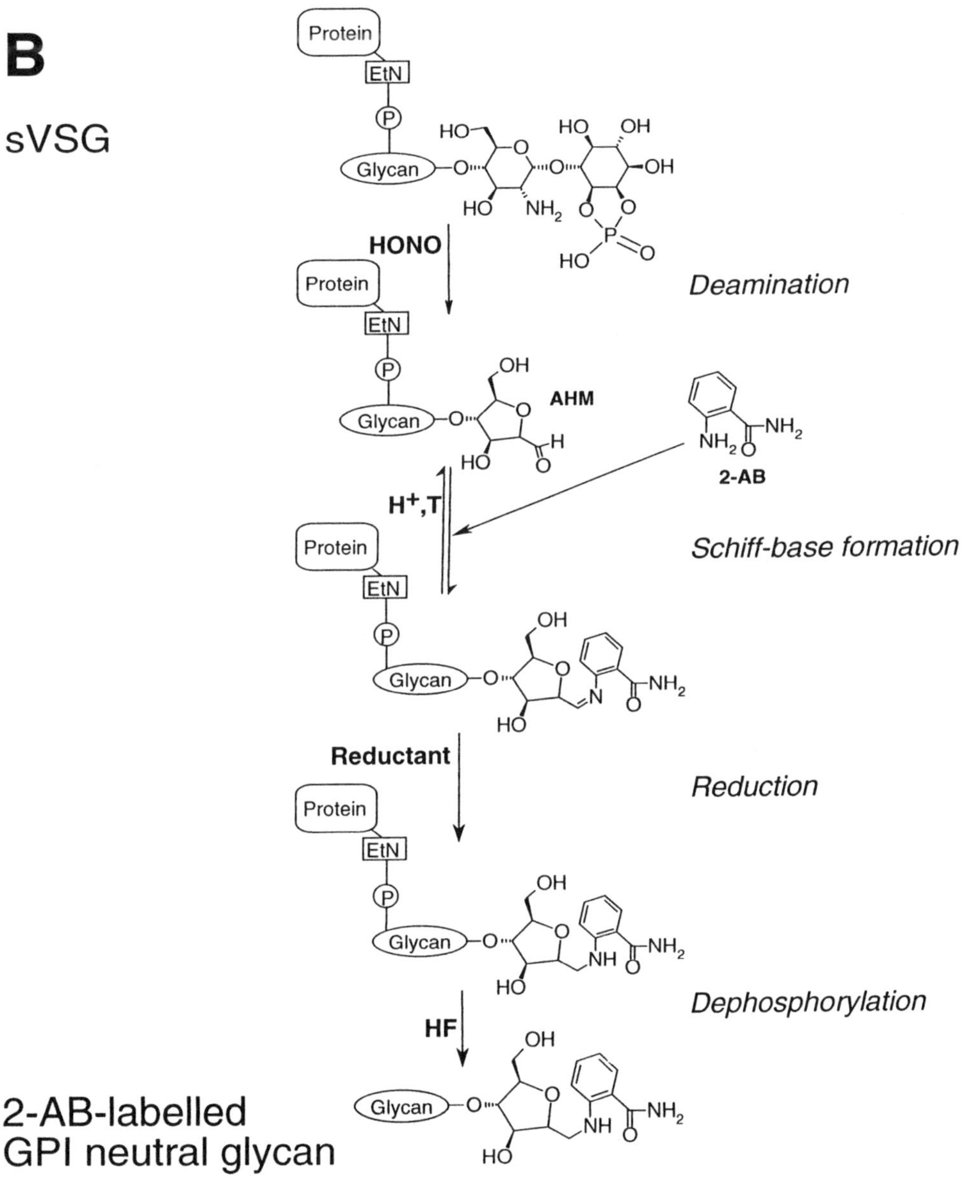

Fig. 2. *(continued)* **(B)** The reaction scheme for producing 2-AB-labeled GPI neutral glycans from sVSG, or any other GPI-anchored glycoprotein, using glycoprotein on a PVDF membrane; the on-the-blot method. In both procedures, dephosphorylation with cold aqueous HF is used to release the GPI glycan, either before (in the tube) or after (on the blot) fluorescent labeling. The fluorescent labeling procedure relies on the conversion of the GPI GlcN residue to 2,5-anhydromannose (AHM) by nitrous acid (HONO) deamination, followed by reductive amination with 2-AB.

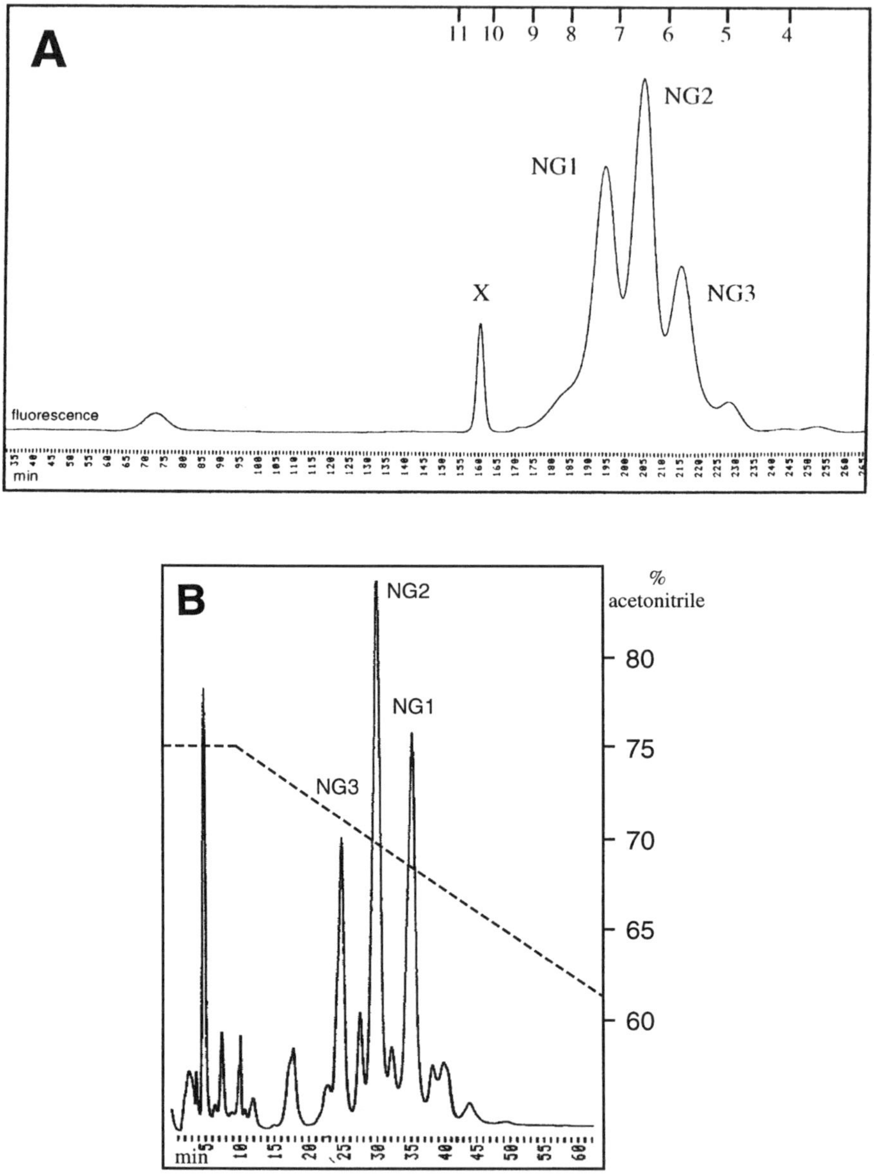

Fig. 3. Chromatography and identification of 2-AB-labeled GPI neutral glycans prepared from purified *T. brucei* VSG. A sample of 0.4 mg (7 nmol) VSG117 was treated as described in **Subheading 3.1.** **(A)** A fraction (1 nmol) of the resulting 2-AB-labeled GPI neutral glycans was analyzed by Bio-Gel P4 gel filtration. The peak marked "X" is a reagent artifact. **(B)** Another fraction (1 nmol) was analyzed by normal-phase HPLC on a 4.6 × 200 mm PolyGlycoPlex column. *(continued)*

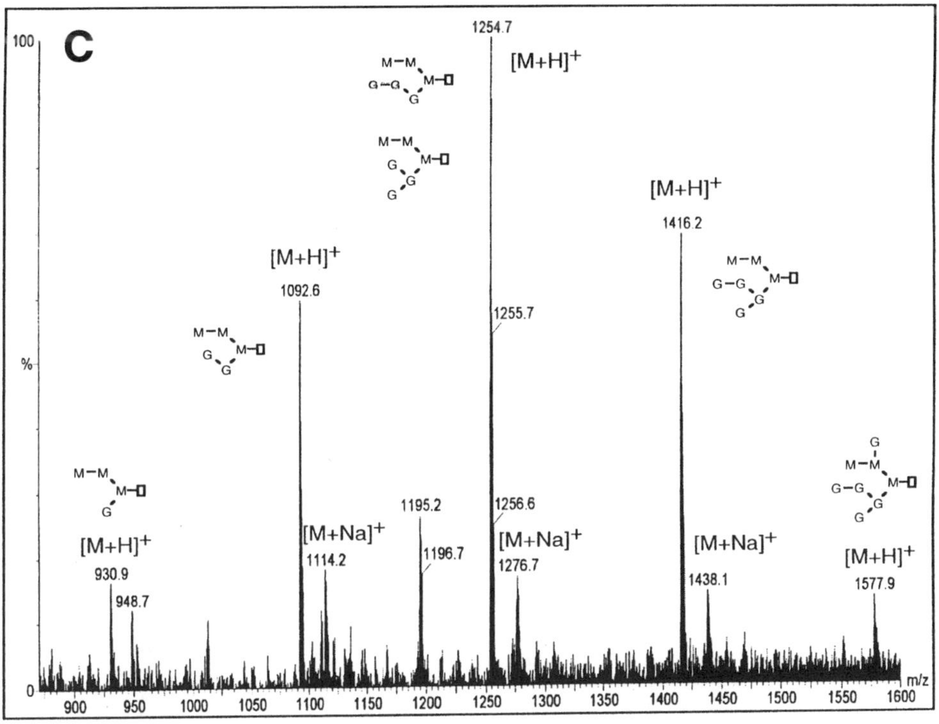

Fig. 3. *(continued)* **(C)** A positive-ion electrospray mass spectrum of 30 pmol of the 2-AB-labeled GPI neutral glycans in methanol–acetic acid–water (1:1:1, v/v/v) confirmed that the expected 2-AB derivatives had been made. *(continued on next page)*

18. Chromatography tank with rack (Oxford GlycoSystems).
19. Butan-1-ol–ethanol–water mixed in the ratio 4:1:1, by volume. The paper chromatography tank should be lined with 3MM Whatman paper, with some of the solvent in the bottom.
20. Access to a fume cupboard.
21. Longwave UV lamp (e.g., BDH-Merck).
22. Access to a microcentrifuge (e.g., Eppendorf microcentrifuge 5415C, BDH-Merck).
23. Microcentrifuge filters (Sigma, Poole, UK).
24. Access to a Bio-Gel P4 system (e.g., the RAAM 1000 or 2000 GlycoSequencer system, Oxford GlycoSystems).
25. 30% Acetic acid in water (BDH-Merck).
26. Acetonitrile, Aristar-grade (BDH).
27. Access to an SDS-PAGE system (e.g., Sigma).
28. PVDF membrane (Amersham, Buckinghamshire, UK).
29. Access to a blotting apparatus (e.g., a semidry blotting apparatus from Hoefer (Pharmacia Biotech Ltd., Little Chalfont, UK).

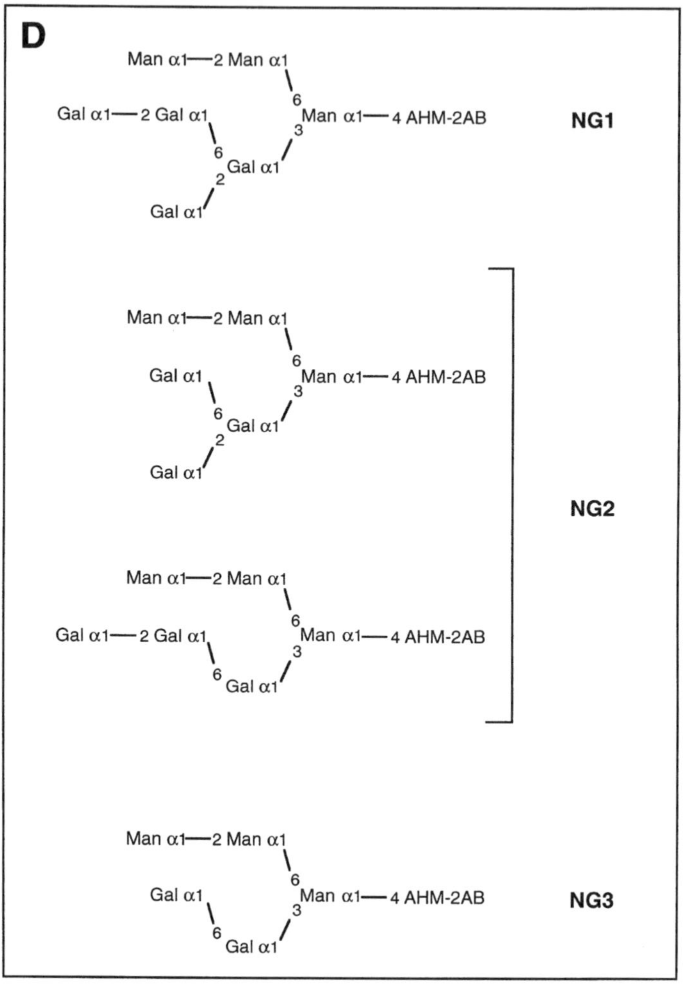

Fig. 3. *(continued)* (D) The structures of neutral glycan fractions NG1, NG2, and NG3.

30. Amido-black (Sigma).
31. Razor blade.
32. Fluorescence detector (e.g., Gilson Model 121; Anachem Ltd., Luton, UK).
33. PolyGlycoPlex column, 4.6 × 200 mm and a PolyGlycoPlex microbore column, 1 × 150 mm (Hichrom, Reading, UK).
34. Access to a Microbore HPLC system, e.g., an Ultrafast Microprotein Analyzer (Michrom Bioresources, CA).
35. Liquid nitrogen.
36. Powder-free gloves (BDH-Merck).
37. Dextran grade C (BDH-Merck).

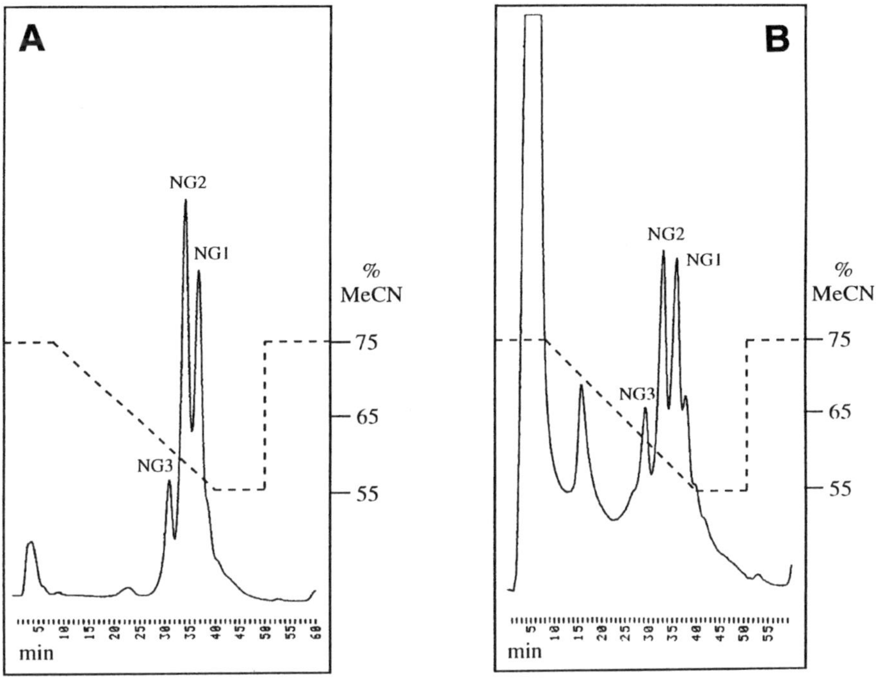

Fig. 4. Normal-phase microbore HPLC of 2-AB-labeled GPI neutral glycans prepared from *T. brucei* VSG. **(A)** An aliquot of the pooled peaks from the Bio-Gel P4 chromatogram (**Fig. 3A**), corresponding to 60 pmol of purified 2-AB-labeled GPI neutral glycans was used to calibrate the 1 × 150 mm microbore PolyGlycoPlex column. **(B)** A Western blot band of 5 μg (100 pmol) VSG117 was treated as described in **Subheading 3.3.**, and the resulting 2-AB-labeled GPI neutral glycans were analyzed by normal-phase microbore HPLC.

3. Methods

3.1. 2-AB Labeling of a GPI-Anchored Protein in the Tube

1. Freeze-dry essentially salt- and detergent-free samples of GPI-anchored protein in screw top Eppendorf tubes.
2. Add 50 μL ice-cold 48% aqueous hydrofluoric acid (aq. HF) to each tube, and dephosphorylate the samples by leaving them for 60–72 h on ice water in a Dewar flask.
3. Remove the HF by freeze-drying (*see* **Note 1**).
4. Add 100 μL water to each tube and freeze-dry again. Repeat this step twice.
5. Resuspend the samples in 30 μL 0.3 *M* sodium acetate, pH 4.0, with the aid of a sonicating water bath.
6. Add 15 μL freshly prepared 1 *M* sodium nitrite solution and allow the nitrous acid deamination to continue for 2 h at 37°C (*see* **Note 2**).

7. Add 100 μL water to each sample, and apply them to C8 Isolute cartridges (preactivated with 5 column volumes of water, followed by 5 column volumes of methanol and 5 column volumes of water). Elute GPI neutral anchor glycans with 600 μL of water (*see* **Note 3**).
8. Desalt samples by passage through 0.2 mL AG50X12 (H^+) over 0.2 mL AG3X4 (OH^-). Elute with 1.5 mL of water.
9. Dry the eluate in a Speed-Vac or rotary evaporator. Redissolve samples in 100 μL water, transfer them to 0.5-mL microtubes, and dry them in a Speed-Vac (*see* **Note 4**).
10. Label the samples with 2-aminobenzamide, using the reagents supplied with the signal 2-AB labeling kit (Oxford GlycoSystems), following the manufacturer's instructions.
11. Add 150 μL of vial C (30% glacial acetic acid) to vial B (350 μL of DMSO), and mix by pipet action (*see* **Note 5**).
12. Add 100 μL of this mixture to vial A (0.35 *M* 2-AB dye), and mix until the dye is dissolved.
13. Add the entire contents of vial A to vial D (1 *M* $NaCNBH_4$), and mix by pipet action until the reductant is completely dissolved.
14. Add 5 μL of the final labeling reagent to each dried glycan sample. Cap the microtubes, mix thoroughly, and tap to ensure that the labeling solution is at the bottom of the tube.
15. Place the tubes in a heating block set at 65°C and incubate for 2 h. Do not use a water bath. To encourage complete dissolution, vortex the samples 30 min after the start of the incubation.
16. Remove samples from the heating block, centrifuge briefly, and allow to cool to room temperature.

3.2. Cleanup Procedures

One of the following procedures can be used for cleaning up the samples after labeling.

3.2.1. Ascending Paper Chromatography

1. Spot the entire reaction mixture in a single transfer to a 3MM Whatman paper strip and let it dry for 30 min at room temperature.
2. Place the strip into a rack and transfer it to the pre-equilibrated chromatography tank filled with 30 mL of BuOH–EtOH–water (4:1:1 v/v/v).
3. After about 60 min, when the solvent front is within 1 cm of the end of the paper, remove the strip from the tank, and dry in a fume cupboard.
4. Localize the fluorescence on the paper strip using a longwave UV lamp in the darkroom (*see* **Note 6**).
5. Cut the origin from the strip, place it into a prewetted 0.22-μm microcentrifuge filter (*see* **Note 7**), soak for 10 min with 0.5 mL of water, then centrifuge for 10 min at 3400*g*. Repeat the last two steps twice, and dry the resulting 1.5 mL of water.

6. Redissolve the sample in 100 µL water, and apply onto the Bio-Gel P4 column. If another column is to be used, redissolve the sample in the appropriate solvent.

3.2.2. OGS Cleanup Cartridge

These cartridges are supplied with the signal 2-AB labeling kit (Oxford GlycoSystems) and are used following the manufacturer's instructions.

1. Prepare cartridges by washing with 1 mL water, followed by 1 mL 30% acetic acid in water and 1 mL acetonitrile.
2. Spot labeled samples onto the center of the wet disk and leave for 15 min.
3. Rinse the sample vial again with 100 µL acetonitrile and apply to the disk.
4. Wash the disk with 1 mL acetonitrile, followed by 5 × 1 mL 4% water in acetonitrile.
5. Place the cartridge over a 2-mL Eppendorf tube and recover the glycans by eluting with 3 × 0.5 mL of water.
6. Filter the sample through a prewetted 0.22-µm microcentrifuge filter, dry, and redissolve in water or solvent for further analysis.

3.3. Small Scale 2-AB-Labeling on the Blot

1. Apply 5 µg (or an equivalent of at least 100 pmols) on a 10% polyacrylamide gel, and subject to SDS-PAGE (*see* **Note 8**).
2. Transfer proteins from the gel to a PVDF membrane (*see* **Note 9**) by Western blotting (*see* **Note 10**).
3. Stain the PVDF membrane with amido-black, cut out the protein bands of interest using a razor blade, and transfer into screw-top Eppendorf tubes (*see* **Note 11**).
4. Deaminate the samples by completely submerging the blot strips in 50 µL of 0.3 *M* NaAc, pH 4.0 and 50 µL freshly prepared 1 *M* sodium nitrite, and leave for 2 h at 37°C.
5. Wash the strips 3× with water to remove salt, transfer into 0.5-mL Eppendorf tubes, and dry.
6. Prepare 2-AB labeling reagent as described above (**Subheading 3.1., steps 10–16**).
7. Take care to completely wet each blot strip with the labeling reagent (usually 15 µL are sufficient) (*see* **Note 12**), cap the tubes, and label the strips for 2–3 h at 65°C in a heating block.
8. Wash the blot strips 3× with about 10 mL 50% acetonitrile, transfer to screw top Eppendorf tubes and dry.
9. Add 40 µL (or as much as needed to submerge the strip) of ice-cold 48% aqueous HF, and dephosphorylate the samples by leaving them for 60–72 h on ice water.
10. Remove the HF by freeze-drying (*see* **Note 1**).
11. Add 100 µL water to each tube and freeze-dry again. Repeat this step.
12. Recover the released 2-AB-labeled neutral glycans by washing the blot strips and Eppendorf tubes several times, using a total of 500 µL water for each sample.

13. Filter the 500 μL containing the glycans through a prewashed 0.22-μm microcentrifuge filter, and dry.
14. Redissolve in 100 μL water for application on the Bio-Gel P4 column, or in the appropriate solvent for analysis on other columns (*see* **Note 2**).

3.4. Column Chromatography

3.4.1. Bio-Gel P4 Gel-Filtration Chromatography Using RAAM 1000 Glycosequencer System (Oxford GlycoSystems)

1. Filter all samples through a 0.22-μm microcentrifuge filter, and redissolve in 100 μL water before application to the Bio-Gel P4 gel-filtration system.
2. Disconnect the solid-cell radioactivity monitor of the instrument and connect an on-line fluorescence detector fitted with suitable excitation and emission filters (for 2-AB-labeled glycans: $\lambda_{max\ exc}$= 330 nm; $\lambda_{max\ em}$= 420 nm) (*see* **Note 13**). Run 2-AB-labeled glucose oligomer standard (*see* **Note 14**) at constant flow (total run volume 38 mL, total run time 5 h 4 min, flow rate 125 μL/min).
3. Run 2-AB labeled samples using the same program, and record the hydrodynamic volume of the glycans in glucose units (*see* **Note 15**).

3.4.2. Normal-Phase Chromatography Using PolyGlycoPlex Columns

1. Dry 2-AB labeled glycans, redissolve in 500 μL of 75% acetonitrile, and apply to a PolyGlycoPlex column (4.6 × 200 mm). Elute samples at 1 mL/min using a gradient from 75 to 50% acetonitrile. Monitor the eluant using an on-line fluorescence detector. Optimize the gradient conditions to achieve best resolution of peaks.
2. If the amount of 2-AB-labeled sample is very small (<0.1 nmols), e.g., after release of labeled glycans from a blot, use microbore columns (e.g., a 1 × 150 mm PolyGlycoPlex microbore column; Hichrom) for analysis. Dry the sample, redissolve in 100 μL of 75% acetonitrile, and inject onto the column. Elute using a gradient from 75 to 50% acetonitrile, at a flow rate of 50 μL/min. Optimize the gradient conditions to achieve best resolution of peaks.

3.5. Exoglycosidase Digestion and Repurification of 2-AB-Labeled Glycans Using an NH_2 Cartridge

1. Analyze 2-AB-labeled glycans using any of the described chromatographic methods, pool fractions of interest, and dry.
2. Dissolve samples in enzyme-containing buffers ***(48,49)*** and digest for 16 h at 37°C.
3. Inactivate enzymes by heating to 100°C for 5 min.
4. Prepare the NH_2 cartridge by precycling it with 1 mL 80% acetonitrile, followed by 1 mL 40% acetonitrile and 1 mL 80% acetonitrile.
5. Adjust samples to a final concentration of 80% acetonitrile, centrifuge, and apply the supernatant to the cartridge. Leave for 10 min.

6. Wash the cartridge with 5 × 1 mL of 80% acetonitrile.
7. Elute 2-AB-labeled neutral glycans using 4 column volumes of 40% acetonitrile.
8. Dry the sample, redissolve in 100 µL water and filter through a prewetted 0.22 µm microcentrifuge filter before reanalyzing using the same method as before the enzyme treatment.

4. Notes

1. Liquid nitrogen can be used to freeze the samples before putting them onto the freeze-dryer. During the freeze-drying process, the temperature rises above the melting point of aq. HF (–83°C). Although the sample thaws, the liquid stays in the bottom of the tubes, and is not sucked into the vacuum source. It is advisable to clean the freeze-dryer afterwards, because HF is highly corrosive.
2. The nitrous acid is generated *in situ* by the action of weak acid on sodium nitrite. The samples often turn yellow or brown during the deamination procedure.
3. Lipids and proteins (released by the action of HF) stick to the C8 Isolute cartridges; the neutral glycans are recovered in the flow-through.
4. Prior to labeling with 2-AB, every care should be taken not to contaminate the samples with carbohydrates or other aldehyde-containing substances. Always wear powder-free gloves, and protect the underivatized samples as much as possible.
5. If the DMSO is frozen, warm it gently to between 30°C and 65°C in a heating block, before opening the vial.
6. 2-AB-labeled neutral glycans will be found at or near the origin; unused label migrates further up.
7. Microcentrifuge filters must be prepared by adding 2 × 400 µL of water and centrifuging at 3400*g* before applying the sample.
8. Both normal-sized and minigels (8 × 10 cm) can be used: The latter is preferable, because less labeling solution will be needed to cover the resulting blot strips.
9. Unlike nitrocellulose, PVDF membranes withstand HF treatment. They must be soaked in MeOH for 1 min before use.
10. The gel is blotted onto a PVDF membrane in a semidry blotting apparatus, using 40 mA per gel for 1 h. Wet blotting may be used instead. It is worth checking the transfer efficiency by Coomassie-staining the polyacrylamide gel afterwards, to be sure that all the material was transferred.
11. The blot strips can be destained using water. However, destaining is not necessary, because the amido-black stain has no adverse effect on the labeling procedure.
12. If the blot strips were not destained, they will turn dark during this step. If they were destained, they become translucent.
13. If you have access to a RAAM 2000 instrument with an built-in fluorescence detector, this step is not necessary.
14. The glucose oligomer internal standards are prepared by partial acid hydrolysis of dextran. 200 mg of dextran is dissolved in 1 mL 0.1 *M* HCl and hydrolyzed at 100°C for 3 h. The acid is removed by passage through 0.5 mL AG3X4(OH^-) and elution with 3 mL of water. The resulting solution (50 mg/mL) contains a series of glucose oligomers from Glc_1 to about Glc_{30} and is stored at –20°C.

15. The GPI neutral glycan hydrodynamic volumes are expressed in glucose units (GU) by linear interpolation of the elution position of the sample glycan peaks between adjacent glucose oligomer external standards.

Acknowledgments

This work was supported by the Wellcome Trust. N. Z. thanks the Wellcome Trust for a PhD studentship. M. A. J. F. is a Howard Hughes Medical Institute International Research Scholar. Thanks to David Neville and Angela Mehlert for suggestions and assistance with this work.

References

1. Ferguson, M. A. J., Homans, S. W., Dwek, R. A., and Rademacher, T. W. (1988) Glycosyl-phosphatidylinositol moiety that anchors *Trypanosoma brucei* Variant surface glycoprotein to the membrane. *Science* **239,** 753–759.
2. Gerold, P., Striepen, B., Reitter, B., Geyer, H., Geyer, R., Reinwald, E., Risse, H. J., and Schwarz, R. T. (1996) Glycosyl-phosphatidylinositols of *Trypanosoma congolense:* two common precursors but a new protein-anchor. *J. Mol. Biol.* **261,** 181–194.
3. Field, M. C., Menon, A. K., and Cross, G. A. M. (1991) A glycosylphosphatidylinositol protein anchor from procyclic stage *Trypanosoma brucei:* lipid structure and biosynthesis. *EMBO J.* **10,** 2731–2739.
4. Ferguson, M. A. J., Murray, P., Rutherford, H., and McConville, M. J. (1993) A simple purification of procyclic acidic repetitive protein and demonstration of a sialylated glycosyl-phosphatidylinositol membrane anchor. *Biochem. J.* **291,** 51–55.
5. Bütikofer, P. and Boschung, M. (1995) Glycosyl inositolphospholipid-anchored structures in *Herpetomonas davidi. Mol. Biochem. Parasitol.* **74,** 65–75.
6. Güther, M.L.S., Almeida, M. L. C, Yoshida, N., and Ferguson, M. A. J. (1992) Structural studies on the glycosylphosphatidylinositol membrane anchor of *Trypanosoma cruzi* 1G7-Antigen. The structure of the glycan core. *J. Biol. Chem.* **267,** 6820–6828.
7. Heise, N., Almeida, M. L. C., and Ferguson, M. A. J. (1995) Characterisation of the lipid moiety of the glycosylphosphatidylinositol anchor of *Trypanosoma cruzi* 1G7-antigen. *Mol. Biochem. Parasitol.* **70,** 71–84.
8. Bertello, L. E., Andrews, N. W., and Lederkremer, R. M. (1996) Developmentally regulated expression of ceramide in *Trypanosoma cruzi. Mol. Biochem. Parasitol.* **79,** 143–151.
9. Couto, A.S., Lederkremer, R. M., Colli, W., and Alves, M. J. M. (1993) The glycosylphosphatidylinositol anchor of the trypomastigote-specific Tc-85 glycoprotein from *Trypanosoma cruzi*. Metabolic labeling and structural studies. *Eur. J. Biochem.* **217,** 597–602.
10. Abuin, G., Couto, A. S., de Lederkremer, R. M., Casal, O. L., Galli, C., Colli, W., and Alves, M. J. M. (1996) *Trypanosoma cruzi:* the TC-85 surface glycoprotein shed by trypomastigotes bears a modified glycosylphosphatidylinositol anchor. *Exp. Parasitol.* **82,** 290–297.

11. Previato, J.O., Jones, C., Xavier, M. T., Wait, R., Travassos, L. R., Parodi, A. J., and Mendoça-Previato, L. (1995) Structural characterisation of the major glycosylphosphatidylinositol membrane anchored glycoprotein from epimastigote forms of *Trypanosoma cruzi* strains. *J. Biol. Chem.* **270,** 7241–7250.
12. Serrano, A. A., Schenkman, S., Yoshida, N., Mehlert, A., Richardson, J. M., and Ferguson, M. A. J. (1995) The lipid structure of the glycosylphosphatidylinositol-anchored mucin-like sialic acid acceptors of *Trypanosoma cruzi* changes during parasite differentiation from epimastigotes to infective metacyclic trypomastigote forms. *J. Biol. Chem.* **270,** 27,244–27,253.
13. Schneider, P., Ferguson, M. A. J., McConville, M. J., Mehlert, A., Homans, S. W., and Bordier, C. (1990) Structure of the glycosyl-phosphatidylinositol membrane anchor of the *Leishmania major* promastigote surface protease. *J. Biol. Chem.* **265,** 16,955–16,964.
14. McConville, M.J., Collidge, T. A. C., Ferguson, M. A. J., and Schneider, P. (1993) The glycoinositol phospholipids of *Leishmania mexicanal* promastigotes. Evidence for the presence of three distinct pathways of glycolipid biosynthesis. *J. Biol. Chem.* **268,** 15,595–15,604.
15. Tomavo, S., Dubremetz, J.-F., and Schwarz, R. T. (1993) Structural analysis of glycosyl-phosphatidylinositol membrane anchor of the *Toxoplasma gondii* tachyzoite surface glycoprotein gp23. *Biol. Cell* **78,** 155–162.
16. Gerold, P., Schofield, L., Blackman, M. J., Holder, A. A., and Schwarz, R. T. (1996) Structural analysis of the glycosyl-phosphatidylinositol membrane anchor of the merozoite surface proteins-1 and -2 of *Plasmodium falciparum. Mol. Biochem. Parasitol.* **75,** 131–143.
17. Azzouz, N., Striepen, B., Gerold, P., Capdeville, Y., and Schwarz, R. T. (1995) Glycosylinositol-phophoceramide in the free-living protozoan *Paramecium primaurelia:* modification of core glycans by mannosyl phosphate. *EMBO J.* **14,** 4422–4433.
18. Fankhauser, C., Homans, S. W., Thomas-Oates, J. E., McConville, M. J., Desponds, C., Conzelmann, A., and Ferguson, M. A. J. (1993) Structures of glycosylphosphatidylinositol membrane anchors from *Saccharomyces cerevisiae. J. Biol. Chem.* **268,** 26365–26374.
19. Haynes, P.A., Gooley, A. A., Ferguson, M. A. J., Redmond, J. W., and Williams, K. L. (1993) Post-translational modifications of the *Dictyostelium discoideum* glycoprotein PsA. Glycosylphosphatidylinositol membrane anchor and composition of O-linked oligosaccharides. *Eur. J. Biochem.* **216,** 729–737.
20. Bütikofer, P., Kuypers, F. A., Shackleton, C., Brodbeck, U., and Stieger, S. (1990) Molecular species analysis of the glycosylphosphatidylinositol anchor of *Torpedo marmorata* acetylcholinesterase. *J. Biol. Chem.* **265,** 18,983–18,987.
21. Mehlert, A., Varon, L., Silman, I., Homans, S. W., and Ferguson, M. A. J. (1993) Structure of the glycosyl-phosphatidylinositol membrane anchor of acetylcholinesterase from the electric organ of the electric fish, *Torpedo californica. Biochem. J.* **296,** 473–479.
22. Homans, S.W., Ferguson, M. A. J., Dwek, R. A., Rademacher, T. W., Anand, R., and Williams, A. F. (1988) Complete structure of the glycosyl phosphatidylinositol membrane anchor of rat brain Thy-1 glycoprotein. *Nature* **333,** 269–272.

23. Stahl, N., Baldwin, M. A., Hecker, R., Pan, K.-M., Burlingame, A. L., and Prusiner, S. B. (1992) Glycosylinositol phospholipid anchors of the scrapie and cellular prion proteins contain sialic acid. *Biochemistry* **31,** 5043–5053.
24. Mukasa, R., Umeda, M., Endo, T., Kobata, A., and Inoue, K. (1995) Characterisation of glycosylphosphatidylinositol (GPI)-anchored NCAM on mouse skeletal muscle cell line C2C12: the structure of the GPI glycan and release during myogenesis. *Arch. Biochem. Biophys.* **318,** 182–190.
25. Taguchi, R., Hamakawa, N., Harada-Nishida, M., Fukui, T., Nojima, K., and Ikezawa, H. (1994) Microheterogeneity in glycosylphosphatidylinositol anchor structures of bovine liver 5'-nucleotidase. *Biochemistry* **33,** 1017–1022.
26. Haas, R., Jackson, B. C., Reinhold, B., Foster, J. D., and Rosenberry, T. L. (1996) Glycoinositol phospholipid anchor and protein C-terminus of bovine erythrocyte acetlcholinesterase: analysis by mass spectrometry and by protein and DNA sequencing. *Biochem. J.* **314,** 817–825.
27. Brewis, I. A., Ferguson, M. A. J., Mehlert, A., Turner, A. J., and Hooper, N. M. (1995) Structures of the glycosyl-phosphatidylinositol anchors of porcine and human renal membrane dipeptidase. Comprehensive structural studies on the porcine anchor and interspecies comparison of the glycan core structures. *J. Biol. Chem.* **270,** 22,946–22,956.
28. Deeg, M. A., Humphrey, D. R., Yang, S. H., Ferguson, T. R., Reinhold, V. N., and Rosenberry, T. L. (1992) Glycan components in the glycoinositol phospholipid anchor of human erythrocyte acetylcholinesterase. *J. Biol. Chem.* **267,** 18,573–18,580.
29. Redman, C. A., Thomas-Oates, J. E., Ogata, S., Ikehara, Y., and Ferguson, M. A. J. (1994) Structure of the glycosylphosphatidylinositol membrane anchor of human placental alkaline phosphatase. *Biochem. J.* **302,** 861–865.
30. Treumann, A., Lifely, M. R., Schneider, P., and Ferguson, M. A. J. (1995) Primary structure of CD52. *J. Biol. Chem.* **270,** 6088–6099.
31. Nakano, Y., Noda, K., Endo, T., Kobata, A., and Tomita, M. (1994) Structural study on the glycosyl-phosphatidylinositol anchor and the asparagine-linked sugar chain of a soluble form of CD59 in human urine. *Arch. Biochem. Biophys.* **311,** 117–126.
32. Sugita, Y., Nakano, Y., Oda, E., Noda, K., Tobe, T., Miura, N.-H., and Tomita, M. (1993) Determination of carboxyl-terminal ersidue and disulfide bonds of MACIF (CD59), a glycosyl-phosphatidylinositol-anchored membrane protein. *J. Biochem.* **114,** 473–477.
33. Meri, S., Lehto, T., Sutton, C. W., Tyynelä, J., and Baumann, M. (1996) Structural composition and functional characterization of soluble CD59: heterogeneity of the oligosaccharide and glycosylphosphatidylinositol (GPI) anchor revealed by laser-desorption mass spectrometric analysis. *Biochem. J.* **316,** 923–935.
34. Rudd, P. M., Morgan, B. P., Wormald, M. R., Harvey, D. J., van den Berg, C. W., Davies, S. J., Ferguson, M. A. J., and Dwek, R. A. (1997) The glycosylation of the complement regulatory protein, human erythrocyte CD59. *J. Biol. Chem.* **272,** 7229–7244.

35. McConville, M. J. and Ferguson, M. A. J. (1993) The structure, biosynthesis and function of glycosylated phosphatidylinositols in the parasitic protozoa and higher eucaryotes. *Biochem. J.* **294,** 305–324.
36. Ferguson, M. A. J. (1994) What can GPI do for you? Parasitol. *Today* **10,** 48–52.
37. Stevens, V. L. (1995) Biosynthesis of glycosylphosphatidylinositol membrane anchors. *Biochem. J.* **310,** 361–370.
38. Udenfriend, S. and Kokudula, K. (1995) How glycosyl-phosphatidylinositol anchored membrane proteins are made. *Ann. Rev. Biochem.* **64,** 563–591.
39. Takeda, J. and Kinoshita, T. (1995) GPI- anchor biosynthesis. *TIBS* **20,** 367–371.
40. Medof, M. E., Nagarajan, S., and Tykocinski, M. L. (1996) Cell-surface engineering with GPI-anchored proteins. *FASEB J.* **10,** 574–586.
41. Redman, C. A., Schneider, P., Mehlert, A., and Ferguson, M. A. J. (1995) The glycoinositol-phospholipids of *Phytomonas. Biochem. J.* **311,** 495–503.
42. Previato, J. O., Mendonca-Previato, L., Jones, C., and Fournet, B. (1992) Structural characterisation of a novel class of glycophosphosphingolipids from the protozoan *Leptomonas samueli. J. Biol. Chem.* **267,** 24,279–24,286.
43. Routier, F. H., da Silveira, E. X., Wait, R., Jones, C., Previato, J. O., and Mendonca-Previato, L. (1995) Chemical characterisation of glycosylinositol-phospholipids of *Herpetomonas samuelpessoai. Mol. Biochem. Parasitol.* **69,** 61–69.
44. Carreira, J. C., Jones, C., Wait, R., Previato, J. O., and Previato, L. (1996) Structural variation in the glycoinositolphospholipids of different strains of *Trypanosoma cruzi. Glycoconj. J.* **13,** 955–66.
45. Striepen, B., Zinecker, C. F., Damm, J. B., Melgers, P. A., Gerwig, G. J., Koolen, M., et al. (1997) Molecular structure of the "low molecular weight antigen" of *Toxoplasma gondii:* a glucose α1-4 N-acetylgucosamine makes free glycosylphosphatidylinositols highly immunogenic. *J. Mol. Biol.* **266,** 797–813.
46. McConville, M. J., Schnur, L. F., Jaffe, C., and Schneider, P. (1995) Structure of *Leishmania* liposphosphoglycan: inter- and intra-specific polymorphism in Old World species. *Biochem. J.* **310,** 807–818.
47. Treumann, A., Güther, M. L. S., Schneider, P., and Ferguson, M. A. J. (1998) Analysis of the carbohydrate and lipid components of glycosylphosphatidylinositol structures, in *Methods in Molecular Biology: Glycoscience Protocols* (Hounsell, E. F., ed.), Humana, Totowa, NJ, pp. 213–235.
48. Ferguson, M. A. J. (1992) The chemical and enzymic analysis of GPI fine structure, in *Lipid Modification of Proteins: A Practical Approach* (Hooper, N. M. and Turner, A. J., eds.), IRL Oxford University Press, pp.191–230.
49. Schneider, P. and Ferguson, M. A. J. (1995) Microscale analysis of glycosylphosphatidylinositol structures. *Meth. Enzymol.* **250,** 614–630.

8

Analysis of the Lipid Moiety of GPI Anchor in the Yeast *Saccharomyces cerevisiae*

Fulvio Reggiori, Elisabeth Canivenc-Gansel, and Andreas Conzelmann

1. Introduction

1.1. Radiolabeling of Ceramide-Containing GPI Proteins with [^{3}H]dihydrosphingosine

The glycosylphosphatidylinositol (GPI) anchors on mature proteins in the yeast *Saccharomyces cerevisiae* contain two different lipid moieties: diacylglycerols and ceramides *(1,2)*. Most anchors contain a ceramide, but there are proteins, as for example, Gas1p, that only contain a diacylglycerol *(2)*. The ceramide moiety is not present in the free GPI lipids, which are the precursors of the anchors, but it is introduced into the GPI anchors of GPI proteins by a remodeling mechanism *(1,3,4)*.

In sphingolipid biosynthesis, dihydrosphingosine (DHS) is the precursor of phytosphingosine and in consequence of all the ceramides and sphingolipids *(5,6)*. [^{3}H]Dihydrosphingosine ([^{3}H]DHS) specifically radiolabels ceramide-containing GPI proteins. Thus, [^{3}H]DHS represents a new way to label GPI proteins specifically, in addition to other commonly used compounds, such as [^{3}H]*myo*-inositol and [^{3}H]palmitate. Labeling with [^{3}H]DHS has proven to be a convenient way to analyze the lipid moieties present on distinct GPI-anchored proteins and to study the phenomenon of the GPI anchor lipid remodeling *(4)*.

[^{3}H]DHS may also prove useful in the study of the biosynthesis of ceramide-containing GPIs and GPI proteins in other organism, such as *Trypanosoma cruzi (7,8)*, *Dictyostelium discoideum (9)*, and *Leishmania promastigotes (10)*.

[^{3}H]DHS is synthesized by the catalytic reduction of sphingosine with tritiated hydrogen (**Fig. 1**). The products of this reaction are separated by ascend-

From: *Methods in Molecular Biology, Vol. 116: Protein Lipidation Protocols*
Edited by: M. H. Gelb © Humana Press Inc., Totowa, NJ

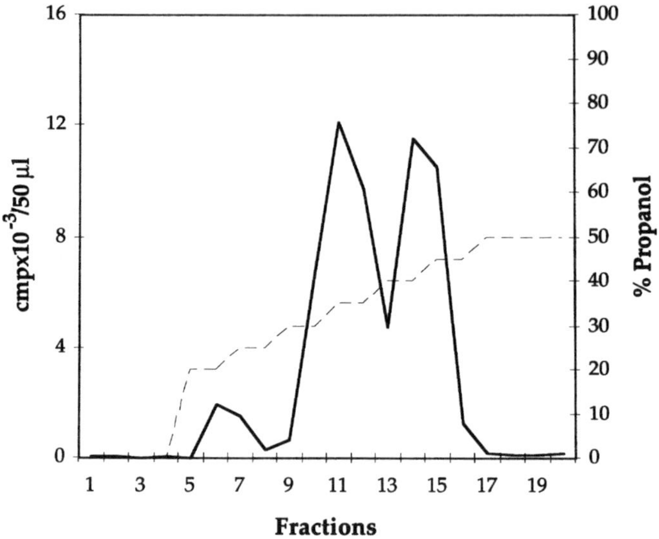

Fig. 1. Palladium-catalyzed reduction of sphingosine by tritiated hydrogen to synthesize [4,5-^{3}H]dihydrosphingosine.

ing TLC, and [^{3}H]DHS is scraped and extracted from the silica gel with methanol ***(11)***.

To radiolabel the yeast *S. cerevisiae* with [^{3}H]DHS, we use salt-dextrose-casamino acid (SDC) medium containing vitamins, but no *myo*-inositol ***(12)***.

This medium should not contain unlabeled DHS, which would be a competitor of the radioactive tracer. [^{3}H]DHS is rapidly absorbed by yeast cells (95–98%), and used for making ceramides and sphingolipids ***(4)***. A small fraction is incorporated into the lipid moiety of GPI anchors.

Yeast cells are broken by glass beads, and most of the lipids are extracted using organic solvents. We generally first use chloroform:methanol:water (10:10:3) to extract long-chain bases (LCB), ceramides, and phospholipids ***(13)***, and then ethanol:water:diethyl ether:pyridine:ammonia (15:15:5:1:0.018), which efficiently extracts sphingolipids ***(14)***. Delipidation of proteins with these two solvents is sufficient to permit the analysis of [^{3}H]DHS radiolabeled proteins using SDS-PAGE ***(15)***. Analysis of the ceramide moiety of GPI anchors requires more rigorous delipidation of proteins in order to eliminate all noncovalently bound lipids.

1.2. Preparation of GPI Anchor Peptides

After extraction with organic solvents, the protein pellet still contains a lot of lipids and other cellular components (cell-wall debris, cytoskeleton). To

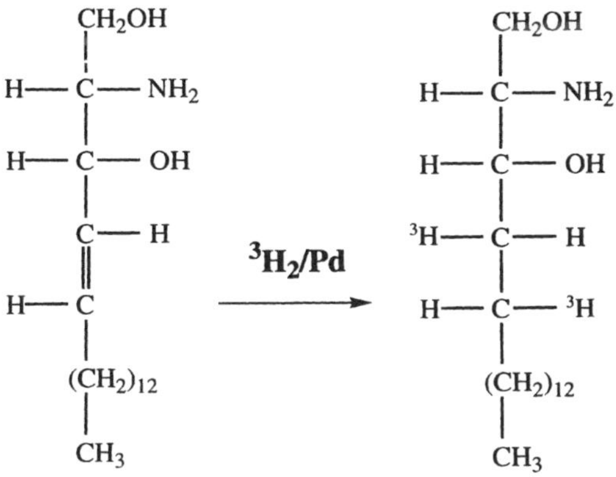

SPHINGOSINE [³H]DIHYDROSPHINGOSINE

Fig. 2. Typical elution profile of GPI peptides from the octyl-Sepharose column.

eliminate these contaminants, two further steps can be performed: affinity chromatography on Concanavalin A (Con A) Sepharose, a lectin with high affinity for mannose, and SDS-PAGE *(1)*. Mannose residues are the prevalent constituent of the *O*- and *N*-glycans of yeast glycoproteins entering the secretory pathway *(17)*. All GPI proteins are glycoproteins and are bound by Con A-Sepharose. When Con A-Sepharose-purified glycoproteins are run on SDS-PAGE, the front usually still contains radioactive lipids. This front can be cut away (1). Subsequent treatment of the gel with pronase permits the cleavage of the GPI proteins into small peptides that can diffuse out of the gel. The peptides bound to a GPI anchor (GPI peptides) can be absorbed on an Octyl-Sepharose column, whereas most other peptides and ions pass through the column **(Fig. 2)** *(3)*. A discontinuous propanol gradient permits the elution of the GPI peptides in a different fraction than the remaining SDS and Triton X-100. Generally, 2.5 OD_{600} of wild-type cells radiolabeled at 30°C with 25 µCi of [^{3}H]DHS yield 50 000 cpm of labeled peptides.

Even when preparing larger amounts of material, it is preferable to radiolabel and extract cells in small aliquots of 2.5 OD_{600}, since small aliquots result in better lipid extraction. Aliquots can be pooled at the moment of Con A-Sepharose chromatography *(4)*.

To obtain GPI peptides from a single GPI-anchored protein, the procedure is similar. In this case, after the separation of protein by SDS-PAGE, the gel is soaked with EN^{3}HANCE (NEN, Hertfordshire, UK) and is processed for fluorography. Individual bands are then cut from the gel and are hydrated *(18)*, and the GPI peptides are extracted with pronase treatment and purified using Octyl-Sepharose chromatography as before.

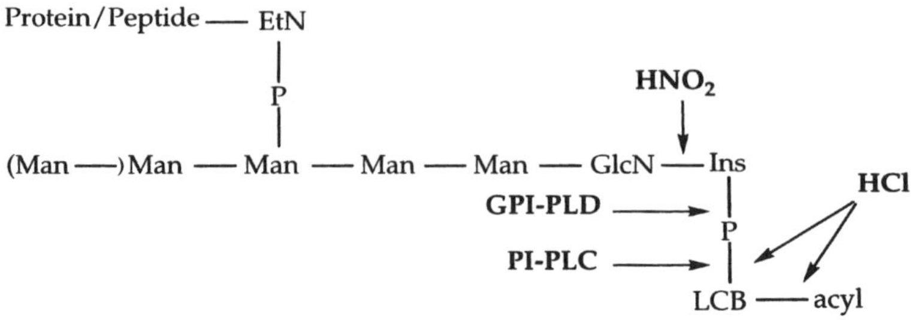

Fig. 3. Specific cleavage of GPI anchors using enzymes and chemical reagents.

1.3. Analysis of the Lipid Moiety of GPI Anchors

To demonstrate that [^{3}H]DHS labeled proteins are GPI-anchored, and the detergent solubilized, delipidated proteins can be treated directly with phosphatidylinositol-specific phospholipase C (PI-PLC) or glycosyl-phosphatidylinositol-specific phospholipase D (GPI-PLD). These enzymes will remove the lipid moiety of GPI anchors *(4)*.

The [^{3}H]DHS-labeled lipid moieties of GPI-anchor peptides can be analyzed using enzymatic and chemical cleavage procedures, and fragments can be analyzed by TLC **(Fig. 3)** *(4)*. LCB, ceramides, phosphoceramides, and inositol-phosphoceramides (IPC) are obtained by strong acid hydrolysis *(1)*, PI-PLC treatment *(4,20)*, GPI-PLD treatment *(21)*, and nitrous acid treatment *(22)*, respectively. Cleavage by nitrous acid is very specific for GPI anchors. After their purification by preparative TLC *(16)*, the more complex ones of these products can further be fragmented by strong acid hydrolysis to liberate the LCB and by phosphatase treatment to transform phosphoceramides into ceramides.

2. Materials

2.1. Purification of [^{3}H]dihydrosphingosine by Preparative TLC

1. [^{3}H]dihydrosphingosine (by special request from NEN, Du Pont De Nemours, Les Ulis, France), SA: 32.3 Ci/mmol.
2. Methanol.
3. TLC tank equilibrated with the solvent chloroform:methanol: 2 *M* ammonium hydroxide ($CHCl_3$:MeOH: 2*M* NH_4OH), (40:10:1 [v/v/v]) (*see* **Note 1**).
4. 20 × 10 cm silica gel 60 TLC plate (0.25-mm thickness) (Merck, Darmstadt, Germany).
5. Two spatulas.
6. Aluminum foil.
7. 50-mL polypropylene tubes (Falcon, Becton Dickinson).

8. Glass pipet.
9. Water bath sonicator.
10. Centrifuge.
11. Berthold scanner LB2842 (Berthold AG, Regensdorf, Switzerland).
12. Ventilated hood.

2.2. Radiolabeling of Yeast Cells

1. [^{3}H]dihydrosphingosine.
2. Sterile SDC medium supplemented with vitamins, but no *myo*-inositol ***(12)***, containing 2% glucose plus 2% Casamino acids (Gibco-BRL, Paisley, Scotland).
3. 0.5 *M* sodium fluoride (NaF).
4. 0.5 *M* sodium azid (NaN_3).
5. Sterile water.
6. 50-mL sterile polypropylene tubes (Becton-Dickinson).
7. Water bath incubator with rotating platform.
8. Centrifuge.
9. Polypropylene microcentrifuge tubes (1.5 mL).
10. Microcentrifuge.
11. Vortex.
12. Water bath sonicator.

2.3. Protein Delipidation for Gel Electrophoresis

1. Chloroform:methanol ($CHCl_3$:MeOH), (1:1 [v/v]).
2. Chloroform:methanol: water ($CHCl_3$:MeOH:H_2O), (10:10:3 [v/v/v]).
3. Ethanol:water:diethylether:pyridine:concentrated aqueous ammonium hydroxide (15:15:5:1:0.018 [v/v/v/v/v]).
4. Water bath sonicator.
5. Glass beads, diameter: 0.45 mm (B. Braun Melsungen AG, Melsungen, Germany).
6. Vortex.
7. Screw-top polypropylene microcentrifuge tubes (1.5 mL).
8. Microcentrifuge.
9. Rotary evaporator with trap cooling 40–104°C (Speed-Vac, Savant, Farmingdale, NY).

2.4. Con A-Sepharose Affinity Chromatography of GPI Proteins

1. Two-times concentrated sample buffer: 4% sodium dodecylsulfate (SDS), 20% glycerol, 100 m*M* Tris-HCl, 10% β-mercaptoethanol, 0.02% bromophenol blue, pH 6.8 (15).
2. Con A-Sepharose buffer: 50 m*M* Tris-HCl, 1% Triton X-100, 150 m*M* NaCl, 1 m*M* $CaCl_2$, 1 m*M* $MgCl_2$, pH 7.4.
3. Con A-Sepharose (Pharmacia, Uppsala, Sweden).
4. 15-mL plastic tubes.

5. Water bath.
6. Vortex.
7. Water bath sonicator.
8. Rotating wheel.

2.5. SDS-PAGE and Pronase Treatment

1. Pronase (Boehringer Mannheim GMBH, Mannheim, Germany).
2. Pronase buffer: 100 m*M* Tris-HCl, 1 m*M* $CaCl_2$, 0.02% Triton X-100, 10 m*M* NaN_3, 20 µg/mL gentamycin, pH 8.0.
3. 0.1% Triton X-100.
4. Razor blade.
5. 2-mL Polypropylene microcentrifuge tubes.
6. Water bath sonicator.
7. Rotating wheel.

2.6. Octyl-Sepharose Affinity Ourification of GPI-Anchor Peptides

1. Octyl-Sepharose (Pharmacia).
2. 20, 25, 30, 35, 40, 45, and 50% propanol (v/v).
3. 5% Propanol in 100 m*M* ammonium acetate, pH 5.5.
4. 10% Acetic acid.
5. 2-mL syringe.
6. Whatman paper 1.
7. Polypropylene microcentrifuge tubes (1.5 mL).

2.7. Preparation of GPI-Anchor Peptides from Individual Proteins

1. Hydration solution: 10% isopropanol, 10% acetic acid, 80% water.
2. Dimethylsulfoxide (DMSO).
3. Fixation solution: 40% methanol, 10% acetic acid, 50% water.
4. EN^3HANCE (NEN, Boston, MA).
5. X-OMAT film (Eastman Kodak C0., Rochester, NY).
6. Razor blade.
7. Plastic tubes.
8. Tweezers.

2.8. PI-PLC and GPI-PLD Treatments

Buffers labeled I are used to treat intact proteins, and buffers labeled II are used for treatment of GPI-anchor peptides and of lipids.

1. GPI-PLD buffer I: 50 m*M* Tris-HCl, 10 m*M* sodium chloride, 2.6 m*M* calcium chloride ($CaCl_2$), 0.018% Triton X-100, pH 4.5.
2. GPI-PLD buffer II: 50 m*M* Tris-HCl, 10 m*M* sodium chloride, 2.6 m*M* calcium chloride ($CaCl_2$), 20% propanol, pH 4.5.
3. GPI-PLD from bovine serum (Boehringer Mannheim GMBH).
4. PI-PLC buffer I: 20 m*M* Tris-HCl, 0.2 m*M* EDTA, 0.1% Triton X-100, pH 7.5.
5. PI-PLC buffer II: 20 m*M* Tris-HCl, 0.2 m*M* EDTA, 20% propanol, pH 7.5.

6. PI-PLC from *B. cereus* (Boehringer Mannheim GMBH).
7. Four-times concentrated sample buffer: 8% SDS, 40% glycerol, 200 m*M* Tris-HCl, 20% β-mercaptoethanol, 0.04% bromophenol blue, pH 6.8.
8. Polypropylene microcentrifuge tubes (1.5 mL).
9. Microcentrifuge.
10. Incubator.

2.9. Nitrous Acid Treatment

1. HNO_2 buffer: 100 m*M* sodium acetate (NaOAc), 0.01% Zwittergent 3-16, pH 3.5.
2. 0.5 *M* sodium nitrite ($NaNO_2$) (*see* **Note 14**).
3. 0.5 *M* sodium chloride (NaCl).
4. Water-saturated *n*-butanol (*see* **Note 15**).
5. *n*-butanol-saturated water.
6. Polypropylene microcentrifuge tubes (1.5 mL).
7. Microcentrifuge.
8. Incubator.
9. Water bath sonicator.

2.10. Strong Acid Hydrolysis

1. 1 *M* hydrochloric acid in methanol (MeOH-HCl) (*see* **Note 16**).
2. Analytical-grade methanol (>99.98%, MeOH).
3. Polypropylene microcentrifuge tube with screw top (1.5 mL).
4. Microcentrifuge.
5. Incubator.

2.11. Alkaline Phosphatase Treatment

1. Alkaline phosphatase buffer: 0.1 *M* glycine, 10 m*M* magnesium chloride ($MgCl_2$), 2 m*M* zinc chloride ($ZnCl_2$), 0.1% Triton X-100, pH 10.1.
2. Alkaline phosphatase from bovine intestine (Sigma Chemical Co., St. Louis, MO).
3. Polypropylene microcentrifuge tubes (1.5 mL).
4. Microcentrifuge.
5. Incubator

3. Methods

3.1. Purification of [^{3}H]dihydrosphingosine by Preparative TLC

1. Process material immediately on arrival to prevent strong radiolysis.
2. Draw a starting line on a TLC plate using a pencil 2 cm from the bottom of a TLC plate.
3. Work in a ventilated hood wearing gloves and a face mask. Ventilation should be strong enough to direct aerosols away from you. Spot [^{3}H]DHS at a width of 5 cm using an Eppendorf micropipet (*see* **Note 2**). Spot 0.2 mg [^{3}H]DHS.cm. Let plate air-dry after each application. Finally air dry for 10 min.
4. Run ascending TLC for approx 1 h at room temperature or until the solvent is about 1 cm from the upper edge of the plate.

5. Visualize the radioactivity using a radioactivity scanner and mark the radio-labeled areas with a pencil. The R_f of [^{3}H]DHS is approx 0.41, whereas degradation products migrate with the front (*see* **Note 3**).
6. Wet the marked area with methanol drops, and scrape the gel with the spatula (*see* **Note 4**).
7. Deposite the silica in a 50-mL polypropylene tube containing 4 mL of methanol. Removing silica from the spatula requires the use of a second spatula.
8. At the end, wash the spatulas with 1 mL of methanol.
9. Close the tube and sonicate for 1 min until the powder becomes finely dispersed. Centrifuge at 3000*g* for 5 min, and transfer the supernatant to a new 50-mL polypropylene tube.
10. Add 1 mL of methanol to the pellet, and sonicate for 1 min. Then centrifuge as above and combine the supernatants.
11. To avoid radiolysis, dilute sample in methanol; [^{3}H]DHS must be stored at –20°C at a concentration no higher than 1–5 mCi/mL.

3.2. Radiolabeling of Yeast Cells

1. Grow the yeast cells overnight in SDC medium supplemented with essential additions if auxotrophic strains are used (*see* **Note 5**). Shake cultures at 250 rpm to have a good oxygenation of the medium.
2. Measure the cell concentration by measuring the absorbance at 600 nm and take aliquots of 2.5 OD_{600} of cells in the exponential phase (*see* **Note 6**).
3. Put cells in sterile 50-mL tubes, centrifuge at 1500*g* for 5 min, and discard the supernatant.
4. Resuspend the cells in 1 ml of sterile water, and vortex. Centrifuge at 1500*g* for 5 min, and discard the supernatant.
5. Resuspend the cells in 250 µL of fresh culture medium to a concentration of 10 OD_{600}/mL.
6. Preincubate cells for 20 min at the labeling temperature in the water bath incubator shaking at 250 rpm.
7. Add 25 µCi of [^{3}H]-DHS directly to the medium and incubate for 40 min (10 µCi/OD_{600} of cells).
8. Add 1 mL of fresh prewarmed medium (1:5 dilution), and continue the incubation for 80 min.
9. Arrest metabolism by adding 25 µL of 0.5 *M* NaF and 25 µL of 0.5 *M* NaN_3 (final concentrations: 10 m*M*) and place tubes on ice.
10. Centrifuge at 1500*g* for 5 min (4°C), and discard the supernatant. Resuspend the cells in 1 mL of ice-cold water, and transfer suspension to a 1.5-mL screw-top microcentrifuge tube.
11. Centrifuge at 1500*g* for 5 min (4°C), and discard the supernatant. Wash the cells once more with 1 mL of ice-cold water in the same way.

3.3. Protein Delipidation for Gel Electrophoresis

1. Resuspend the cell pellet in 400 µL of chloroform-methanol (1:1) by sonication, and add 400 µL of dry packed glass beads.

2. Vortex at top speed 5 times for 1 min with 1-min intervals on ice.
3. Remove supernatant to fresh tube (*see* **Note 7**) and wash the beads twice with 400 µL of chloroform:methanol:water (10:10:3) plus brief sonication. Add washes to first supernatant.
4. Centrifuge at 15,000*g* for 5 min (4°C).
5. Remove the supernatant (*see* **Note 8**) and add another 400 µL of chloroform: methanol:water (10:10:3). Sonicate to resuspend the protein pellet, and let sit on ice for 20 min. Centrifuge at 15,000*g* for 5 min (4°C).
6. Remove the supernatant, and extract the proteins twice more as in **step 5**.
7. Remove the supernatant, and add 400 µL of ethanol:water:diethylether: pyridine:ammonia (15:15:5:1:0.018). Sonicate to resuspend the protein pellet, and incubate for 20 min at 37°C.
8. Centrifuge at 15,000*g* for 5 min, remove the supernatant, and dry the protein pellet using a rotary evaporator. It is not necessary to use nitrogen gas to dry the proteins, because the radiolabeled lipids do not contain double bonds (*see* **Note 9**).

3.4. Concanavalin A-Sepharose Affinity Chromatography of GPI Proteins

In the following, we describe a standard labeling of 2.5 OD_{600} of cells. Quantities have to be multiplied if larger amounts of material are prepared.

1. To the delipidated protein pellet from 2.5 OD_{600}, add 25 µL of water, resuspend first by vortexing, then by sonication.
2. Add 25 µL of sample buffer 2×/2.5 OD_{600} and vortex.
3. Boil for 5 min, and centrifuge at 3000*g* for 5 min.
4. Place the supernatant into a tube, and dilute the supernatant 1:41 by adding 2 mL of Con A-Sepharose buffer.
5. Place solution in a plastic tube that contains the appropriate amount of Con A-Sepharose (25 µL/2.5 OD_{600}) (*see* **Note 10**), and incubate on a rotating wheel at room temperature for 2 h.
6. Centrifuge at 1500*g* for 2 min, and remove the supernatant.
7. Add 1 mL of Con A-Sepharose buffer, and resuspend gently.
8. Let sediment for 15 min. Then centrifuge at 1500*g* for 5 min, and discard the supernatant.
9. Continue to wash the Sepharose in this way until the radioactivity in the supernatant becomes constant (approx at 300–500 cpm/mL).
10. Add 150 µL of two-times concentrated sample buffer/100 µL of Con A-Sepharose, and boil for 5 min.
11. Centrifuge at 1500*g* for 5 min; the glycoproteins are in the supernatant.

3.5. SDS-PAGE and Pronase Treatment

1. Load the proteins eluted from Con A-Sepharose on a 7.5% or a 6–10% gradient gel for SDS-PAGE (8 × 10 cm).

2. After electrophoresis, the support of the gel is removed. Use a razorblade to cut away the stacking gel, the empty lanes, and the bottom of the gel up to 0.5 cm above the dye front. The remaining part of the gel, which contains the proteins of interest, is cut into small pieces. Place pieces into a 2-mL microcentrifuge tube.
3. Dry the gel using a rotary evaporator, and add 1.6 mL of pronase buffer. Let gel hydrate, and sonicate briefly.
4. Bring the pH to 8.0, and add 200 µL of preincubated pronase (*see* **Note 11**) (final concentration: 1 mg/mL). Incubate at 37°C overnight.
5. Vortex, remove the supernatant to a fresh tube, and add 1 mL of 0.1% Triton X-100 to the residual gel pieces. Sonicate for 1 min, and then wash the gel pieces by incubating for 10–15 min on a rotating wheel. Sonicate again for 1 min, and remove the supernatant. Wash the gel pieces in the same way once more.
6. Pool the three supernatants, that contain the GPI peptides.

3.6. Octyl-Sepharose Affinity Purification of GPI Anchor Peptides

1. Adjust the pH of the pronase eluate to a pH of about 4.5 by adding 50-µL aliquots of 10% acetic acid.
2. Apply the sample to an Octyl-Sepharose column (*see* **Note 12**), and collect the run through. Reapply the run-through fraction, and start to collect fractions of 0.5 bed volumes.
3. Then add successively 1 bed vol of 20, 25, 30, 35, 40, and 45% propanol, and finally 2 bed volumes of 50% propanol.
4. Determine the radioactivity in an aliquot from in each fraction, pool the radioactive fractions, and dry in rotary evaporator (*see* **Note 13**). At this stage, all radioactivity is present in GPI peptides.

3.7. Preparation of GPI-Anchor Peptides from Individual Proteins

1. After SDS-PAGE as in **Subheading 3.5.**, proteins are fixed in the gel by shaking in the fixation solution for 15 min.
2. Treat gel with EN^{3}HANCE for 40 min, and then precipitate the fluor in water for 20 min.
3. Dry the gel between two sheets of cellophane in a vacuum gel dryer and expose it to X-OMAT film at –80°C for 1 wk.
4. Develop the film, superimpose it onto the gel, and mark the bands corresponding to the radiolabeled proteins.
5. Cut the marked regions out of the gel, and cut them into small pieces using the razorblade.
6. The pieces are put into a tube, and 3 mL of hydration solution are added. Let the gel hydrate for 30 min. Discard the supernatant and remove the pieces of cellophane.
7. Add 3 mL of DMSO, and let stand on the bench for 45 min in order to extract the EN^{3}HANCE. Discard the solvent, and repeat this manipulation twice.
8. To eliminate DMSO, add 3 mL of hydration solution, incubate for 30 min, and remove supernatant.
9. Dry the gel and continue with pronase treatment and octyl-Sepharose chromatography as described in **Subheading 3.5.**, **steps 3–6** and **Subheading 3.6.**

3.8. PI-PLC and GPI-PLD Treatment of Intact Proteins to Be Analyzed by SDS-PAGE

1. Resuspend the delipidated radiolabeled proteins in 75 μL of PI-PLC buffer I or 75 μL of GPI-PLD buffer I by sonication and vortexing (*see* **Note 17**).
2. Boil for 5 min and centrifuge at 1500*g* for 5 min.
3. Divide the supernatant into two aliquots.
4. Add 0.05 U of PI-PLC or 0.5 U of GPI-PLD to one aliquot.
5. Incubate both aliquots at 37°C overnight.
6. Add 12.5 μL of four times concentrated sample buffer, and boil for 5 min.
7. Spin and load the supernatant onto an SDS-PAGE gel.

3.9. Nitrous Acid Treatment

1. Proceed as described in detail by Güther et al. *(22)*.
2. The ultimate *n*-butanol phase contains the IPCs (*see* **Note 18**).

3.10. PI-PLC Treatment of GPI-Anchor Peptides and Lipids

1. Resuspend the dried GPI peptides (**Subheading 3.6.**) or lipids in 100 μL of PI-PLC buffer II by sonication and vortexing *(20)*. Split sample into two equal aliquots.
2. Add 0.05 U of PI-PLC to one aliquot and incubate both aliquots for 3 h at 37°C.
3. Cool down and add 200 μL of water-saturated *n*-butanol. Vortex and let stand to allow for separation of the phases. Centrifuge at 15,000*g* for 1 min.
4. Remove the *n*-butanol phase and re-extract the water phase with 200 μL of water-saturated *n*-butanol as before.
5. Pool the two *n*-butanol phases, and add 50 μL of *n*-butanol-saturated water.
6. Vortex, let stand for 2 min, and then centrifuge at 15,000*g* for 5 min. Remove *n*-butanol phase to fresh tube.
7. The *n*-butanol phase contains the ceramides (*see* **Note 18**).

3.11. GPI-PLD Treatment of GPI Anchor Peptides and Lipids

1. Resuspend the dried GPI peptides (**Subheading 3.6.**) or lipids in 100 μL of GPI-PLD buffer II by sonication and vortexing. Split sample into two equal aliquots.
2. Add 0.5 U of GPI-PLD to one aliquot, and incubate both aliquots overnight at 37°C.
3. Stop the reaction, and desalt the lipids using water-saturated n-butanol as described before (**Subheading 3.10., steps 3–6**).
4. The n-butanol phase contains the phosphoceramides (*see* **Note 18**).

3.12. Strong Acid Hydrolysis

1. Put material in a 1.5-mL screw top tube and dry.
2. Resuspend in 400 μL of 1 *M* hydrochloric acid in methanol by sonication and vortexing.

3. Incubate tightly closed tube overnight at 80°C.
4. Cool down, and centrifuge at 15,000*g* for 1 min. Dry under nitrogen or in a rotary evaporator.
5. Add 200 µL of dry methanol, vortex, and dry.
6. Repeat **step 5** once more. The tube should contain the dried LCB (*see* **Note 18**).

3.13. Alkaline Phosphatase Treatment

1. Resuspended the dried phosphoceramides in 100 µL of alkaline phosphatase buffer by sonication and vortexing.
2. Add 2 U of alkaline phosphatase, and incubate overnight at 37°C.
3. Stop the reaction, and desalt the lipids using water-saturated n-butanol as described before (**Subheading 3.10., steps 3–6**).
4. The *n*-butanol phase contains the ceramides (*see* **Note 18**).

4. Notes

1. Cover the TLC tank inside with two sheets of Whatman no. 3 paper (Whatman International Ltd., Maidstone, UK). Close tank tightly and let tank pre-equilibrate for at least 3 h.
2. [^{3}H]DHS is generally delivered in methanol and can be spotted directly in this solvent. Cool sample on ice, and leave micropipet several seconds in the solvent in order to cool down. Otherwise, warming of methanol in micropipet will increase vapor pressure and methanol will spill out uncontrollably.
3. To identify the band that corresponds to [^{3}H]DHS, we first run an analytical TLC plate with 200,000 cpm of [^{3}H]DHS spotted next to 100 µg of unlabeled DHS (Sigma Chemical Co.). DHS can be visualized by spraying the plate in a ventilated hood with ninhydrine solution (Merck, Darmstadt, Germany) and subsequent heating at 100°C until a red spot appears ***(16)***.
4. To avoid inhalation of the radioactive powder, you have to work in a ventilated hood or in places without air currents. Place your material onto a plastic tray, and work on disposable aluminum foil. Have a waste can on the same tray. Wear face mask.
5. Inoculate different cultures with different quantities of cells from the preculture to make sure that you will have one culture in the exponential phase.
6. One OD_{600} of cells designates 1 mL of supsension having an absorbance of 1. One OD_{600} of cells corresponds to $1–2 \times 10^7$ cells, depending on the strain. Dilute cell suspension 1:10 for the determination of absorbance at 600 nm.
7. The glass beads can obstruct the tip of the micropipet. To avoid this inconvenience, cut off the tip of a 200-µL tip so that glass beads can fall out.
8. Keep this supernatant and the three following washings if you want to analyze the radiolabeled lipid fraction.
9. At this point, the proteins can be analyzed by SDS-PAGE ***(15)***, but if you radiolabel 2.5 OD_{600} of cells with more than 25 µCi of [^{3}H]DHS, it is better to delipidate further using Con A-Sepharose affinity chromatography (*see* **Subheading 3.4.**).

10. Remove unbound Con A from Con A-Sepharose by prewashing. Place the chosen quantity, e.g. 150 µL of Con A-Sepharose, into a 15-mL tube. Let sediment and discard the supernatant. Add 5 mL of Con A-Sepharose buffer, resuspend, let sediment for 15 min, and then centrifuge at 200*g* for 5 min. Discard the supernatant: The Con A-Sepharose is now ready for use.
11. To remove contaminating glycosidases, we dissolve 9 mg of pronase in 200 µL of pronase buffer, and preincubate for 30 min at 37°C.
12. Fix the 2-mL syringe in vertical position, and remove the piston. Place a round piece of Whatman no. 1 paper at the bottom of the syringe, and wash with 50% propanol. Add 1 mL of octyl-Sepharose for material from 10–15 OD_{600} starting cells, and let sediment. Cover the gel with a round piece of Whatman no. 1 paper, and wash the column with 3 vol of 50% propanol. Then equilibrate the resin with 3 vol of 5% propanol in 100 m*M* ammonium acetate, pH 5.5.
13. We generally obtain three peaks of radioactivity **(Fig. 2)**. An initial, small peak contains the detergents (SDS and Triton X-100). The two other peaks eluting between 30 and 40% propanol are not well separated and correspond to GPI peptides. Anchor peptides in these two peaks contain the same ceramide moieties and the same number of mannose residues on their GPI anchor (4 or 5). Indeed, when labeling *sec12* and *sec18* cells at 37°C, the tracer only gets incorporated into GPI anchors with 4 mannoses ***(19)***, but they still elute in two peaks. We believe that the difference between the two peaks is owing to the amino acids on these anchor peptides. This idea is confirmed by the profiles obtained from purified protein bands each of which contains only one of the two peaks.
14. This solution must be freshly prepared in a ventilated hood.
15. To prepare this solution, take equal volumes of *n*-butanol and distilled water, and mix by shaking. Let stand for 1 h to allow the two phases to separate. The upper phase corresponds to the water-saturated *n*-butanol, and the lower phase to the *n*-butanol-saturated water.
16. To prepare this mixture, mix 1 vol of 10 *M* HCl in water with 9 vol of methanol.
17. These enzymatic treatments should be carried out with proteins that have been delipidated by Con A-Sepharose affinity chromatography (**Subheading 3.4., steps 1–9**). To elute proteins from Con A-Sepharose for enzymatic treatments, add 75 µL of PI-PLC buffer I or 75 µL of GPI-PLD buffer I. Sonicate and boil for 5 min. Sediment Sepharose at 200*g* for 5 min, and take the supernatant. Then continue as described in **Subheading 3.8.**, **step 2**.
18. The resulting lipids can be analyzed by TLC. LCB and ceramides are well separated in the solvent system chloroform:methanol:2 *M* ammonium hydroxide ($CHCl_3$:MeOH:2 *M* NH_4OH) (40:10:1 [v/v/v]) (**Subheading 2.1.**); phosphoceramides using chloroform:methanol:2 *M* acetic acid ($CHCl_3$:MeOH:2 *M* CH_3COOH) (18:10:2 [v/v/v]) and IPCs using chloroform:methanol:0.25% potassium chloride ($CHCl_3$:MeOH:0.25% KCl) (55:45:10 [v/v/v]).

    To find out what type of LCB is present in an IPC, phosphoceramide, or ceramide species resolved by TLC, one can scrape the silica using a razorblade and let the powder fall down onto aluminum foil or clean paper with a fold in the middle. When scraping, one has to tilt the plate with the silica gel toward the bottom, so the falling distance of the powder is minimal. Concentrate the silica gel powder in the fold, and pour it into a 2-mL polypropylene tube using a glass funnel. Add 1 mL of chloroform:methanol:water ($CHCl_3$: MeOH: H_2O) (10:10:3 [v/v/v]) through the funnel. Take care to wash down the remaining powder on the funnel. Sonicate for 1 min until the powder becomes finely dispersed. Centrifuge at 15,000*g* for 1 min, and collect the supernatant. Add 500 μL of solvent to the silica gel, sonicate 1 min, centrifuge at 15 000*g* for 1 min and collect the supernatant. Wash the gel once more in the same way. Pool the three supernatants and dry using a rotary evaporator. Migrate the extracted lipids once more on TLC, and repurify them again in the same way. Dry the lipids, desalt by water-butanol partitioning (*see* **Subheading 3.10., steps 3–6**), and continue by strong acid hydrolysis (*see* **Subheading 3.12.**).

References

1. Conzelmann, A., Puoti, A., Lester, R. L., and Despond, C. (1992) Two different types of lipid moieties are present in glycophosphoinositol-anchored membrane proteins of *Saccharomyces cerevisiae. EMBO J.* **11,** 457–466.
2. Fankhauser, C., Homans, S. W., Thomas Oates, J. E., McConville, M. J., Desponds, C., Conzelmann, A., and Ferguson, M. A. J. (1993) Structures of glycosylphosphatidyl-inositol membrane anchors from *Saccharomyces cerevisiae. J. Biol. Chem.* **268,** 26,365–26,374.
3. Sipos, G., Puoti, A., and Conzelmann, A. (1994) Glycosylphosphatidylinositol membrane anchors in *Saccharomyces cerevisiae*: absence of ceramides from complete precursor glycolipids. *EMBO J.* **13,** 2789–2796.
4. Reggiori, F., Canivenc-Gansel, E., and Conzelmann, A. (1997) Lipid remodeling leads to the introduction and exchange of defined ceramides on GPI proteins in the ER and Golgi of *Saccharomyces cerevisiae. EMBO J.* **16,** 3506–3518.
5. Merill, A.H., Jr. and Jones, D.D. (1990) An update of the enzymology and regulation of sphingomyelin metabolism. *Biochem. Biophys. Acta* **1044,** 1–12.
6. Wells, G.B. and Lester, R.L. (1983) The isolation and characterization of a mutant strain of Saccharomyces cerevisiae that requires a long chain base for growth and for synthesis of phosphosphingolipids. *J. Biol. Chem.* **258,** 10,200–10,203.
7. Güther, M. L., Cardoso de Almeida, M. L., Yoshida, N., and Ferguson, M. A. J. (1992) Structural studies on the glycosylphosphatidylinositol membrane anchor of *Trypanosoma cruzi* 1G7-antigen. *J. Biol. Chem.* **267,** 6820–6828.
8. Heise, N., Raper, J., Buxbaum, L. U., Peranovich, T. M. S., and Cardoso de Almeida, M. L. (1996) Identification of complete precursors for the glycosylphosphatidylinositol protein anchors of *Trypanosoma cruzi. J. Biol. Chem.* **271,** 16,877–16,887.
9. Stadler, J., Keenan, T. W., Bauer, G., and Gerisch, G. (1989) The contact site A glycoprotein of *Dictyostelium discoideum* carries a phospholipid anchor of a novel type. *EMBO J.* **8,** 371–377.

10. McConville, M. J. and Ferguson, M. A. J. (1993) The structure, biosynthesis and function of glycosylated phosphatidylinositols in the parasitic protozoa and higher eucaryotes. *Biochem. J.* **294,** 305–324.
11. Crossman, M. W. and Hirschberg, C. B. (1977) Biosynthesis of phytosphingosine by the rat. *J. Biol. Chem.* **252,** 5815–5819.
12. Wickerham, J.L. (1946) A critical evaluation of the nitrogen assimilation tests commonly used in the classification of yeast. *J. Bacteriol.* **52,** 293–301.
13. Puoti, A., Desponds, C., and Conzelmann, A. (1991) Biosynthesis of mannosylinositolphosphoceramide in *Saccharomyces cerevisiae* is dependent on genes controlling the flow of secretory vesicles from the endoplasmic reticulum to the Golgi. *J. Cell Biol.* **113,** 515–525.
14. Smith, S. W. and Lester, R. L. (1974), Inositol phosphorylceramide, a novel substance and the chief member of a major group of yeast sphingolipids containing a single inositol phosphate. *J. Biol. Chem.* **249,** 3395–3405.
15. Laemmli, U. K. (1970) Cleavage of structural proteins during the assembly of the head of bacteriophage T4. *Nature* **227,** 680–685.
16. Jork, H., Funk, W., Fischer, W., and Wimmer, H. (1990) *Thin-Layer Chromatography, Reagents and Detection Methods*, vol. 1a, VCH-Verlag, Weinheim, pp. 411.
17. Herscovics, A. and Orlean, P. (1993) Glycoprotein biosynthesis in yeast. *FASEB J.* **7,** 540–550.
18. Sefton, B. M., Trowbridge, I. S., Cooper, J. A., and Scolnick, E. M. (1982) The transforming proteins of Rous sarcoma virus, Harvey sarcoma virus and Abelson virus contain tightly bound lipid. *Cell* **31,** 465–474.
19. Sipos, G., Puoti, A., and Conzelmann, A. (1995) Biosynthesis of the side chain of yeast glycosylphosphatidylinositol anchors is operated by novel mannosyltransferases located in the endoplasmic reticulum and the Golgi apparatus. *J. Biol. Chem.* **270,** 19,709–19,715.
20. Puoti, A. and Conzelmann, A. (1993) Characterization of abnormal free glycophosphatidylinositols accumulating in mutant lymphoma cells of classes B, E, F, and H. *J. Biol. Chem.* **268,** 7215–7224.
21. Davitz, M. A., Hom, J., and Schenkman, S. (1989) Purification of a glycosylphosphatidylinositol-specific phospholipase D from human plasma. *J. Biol. Chem.* **264,** 13,760–13,764.
22. Güther, M. L., Masterson, W. J., and Ferguson, M. A. J. (1994) The effects of phenylmethylsulfonyl fluoride on inositol-acylation and fatty acid remodeling in African trypanosomes. *J. Biol. Chem.* **269,** 18,694–18,701.

9

Rapid Identification of Cysteine-Linked Isoprenyl Groups by Metabolic Labeling with [^{3}H]Farnesol and [^{3}H]Geranylgeraniol

Douglas A. Andres, Dean C. Crick, Brian S. Finlin, and Charles J. Waechter

1. Introduction

1.1. Background

The posttranslational modification of proteins by the covalent attachment of farnesyl and geranylgeranyl groups to cysteine residues at or near the C-terminus via a thioether bond is now well established in mammalian cells ***(1–6)***. Most isoprenylated proteins are thought to serve as regulators of cell signaling and membrane trafficking. Farnesylation and geranylgeranylation of the cysteinyl residues have been shown to promote both protein–protein and protein–membrane interactions ***(6–8)***. Isoprenylation, and, in some cases, the subsequent palmitoylation, provide a mechanism for the membrane association of polypeptides, which lack a transmembrane domain, and appear to be prerequisite for their in vivo activity ***(6,9,10)***.

Three distinct protein prenyltransferases catalyzing these modifications have been identified ***(1–5)***. Two geranylgeranyltransferases (GGTases) have been characterized, and are known to modify distinct protein substrates. The Caa*X* GGTase (also known as GGTase-1) geranylgeranylates proteins that end in a CaaL(F) sequence, where C is cysteine, *a* is usually an aliphatic amino acid, and the C-terminal amino acyl group is leucine (L) or phenylalanine (F). Rab GGTase (also known as GGTase-2) catalyzes the attachment of two geranylgeranyl groups to paired carboxyl-terminal cysteines in most members of the Rab family of GTP-binding proteins ***(11)***. These proteins terminate in a Cys-Cys, Cys-*X*-Cys or Cys-Cys-*X*-*X* motif, where *X* is a small hydrophobic amino acid. Another set

From: *Methods in Molecular Biology, Vol. 116: Protein Lipidation Protocols*
Edited by: M. H. Gelb © Humana Press Inc., Totowa, NJ

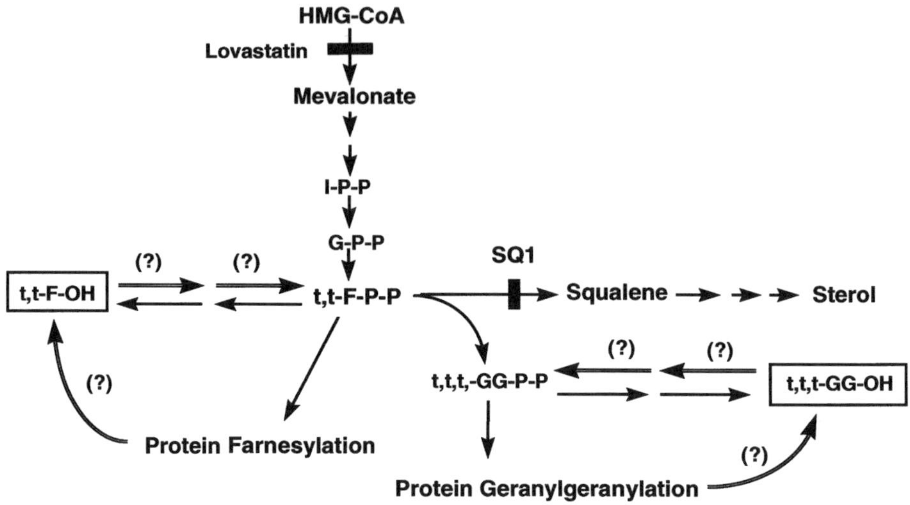

Fig. 1. Proposed salvage pathway for the utilization of F-OH and GG-OH for isoprenoid biosynthesis. Question marks (?) indicate proposed reactions that have not been extensively characterized. Evidence for the presence of microsomal enzymes catalyzing the conversion of F-P-P and GG-P-P to the free isoprenols has been reported by Bansal and Vaidya *(29)*.

of regulatory proteins is modified by protein farnesyltransferase (FTase). All known farnesylated proteins terminate in a tetrapeptide Caa*X* box, wherein C is cysteine, a is an aliphatic amino acid, and *X* has been shown to be a COOH-terminal methionine, serine, glutamine, cysteine, or alanine.

Isoprenylated proteins are commonly studied by metabolic labeling of cultured cells by incubation with [^{3}H]mevalonolactone, which is enzymatically converted to [^{3}H]farnesyl pyrophosphate (F-P-P) and [^{3}H]geranylgeranyl pyrophosphate (GG-P-P) prior to being incorporated into protein **(ref. *12*; Fig. 1)**. Cellular proteins can then be analyzed by gel electrophoresis and autoradiography. However, [^{3}H]mevalonolactone labeling does not directly distinguish between farnesylated and geranylgeranylated proteins. Other methods for the identification of the isoprenyl group attached to protein, which make use of [^{3}H]mevalonolactone labeling, have been described. These methods generally require the isolation of large amounts of labeled protein, extensive proteolysis, a number of column chromatographic steps, and/or cleavage of the thioether bond with Raney nickel or methyl iodide, followed by gas chromatography or HPLC and mass spectrometry to identify the volatile cleavage products (*see* **ref. *1*** and papers cited therein).

The authors have recently shown that mammalian cells can utilize the free isoprenols, GG-OH (geranylgeraniol) and F-OH (farnesol), for the isoprenylation of cellular proteins ***(13,14)***. When C6 glioma cells and green monkey kidney (CV-1) cells were incubated with [^{3}H]F-OH, radioactivity was also incorporated into cholesterol ***(14)***. The observation that the incorporation of label into sterol was blocked by squalestatin 1 (SQ), a potent inhibitor of squalene synthetase ***(15–19)***, suggested that F-OH, and probably GG-OH, are utilized for isoprenoid biosynthesis after being converted to the corresponding activated allylic pyrophosphates, F-P-P and GG-P-P (**Fig. 1**). Preliminary studies have suggested the presence of enzyme systems in mammals and lower organisms that are capable of phosphorylating F-OH and GG-OH ***(20–22)***. Further work is certainly warranted on the isolation and characterization of these enzymes. The developments in understanding the mechanism and physiological significance of the salvage pathway for the utilization of F-OH and GG-OH are discussed in a recent review ***(23)***.

This chapter describes a novel approach for the use of free [^{3}H]farnesol (F-OH) and [^{3}H]geranylgeraniol (GG-OH) in the selective metabolic labeling of farnesylated and geranylgeranylated proteins, respectively, in cultured mammalian cells. In addition, the methods use materials and equipment readily available in most laboratories.

1.2. Experimental Strategy

Utilizing free F-OH or GG-OH as the isotopic precursors has several experimental advantages over metabolic labeling of isoprenylated proteins with mevalonolactone. The isoprenols rapidly enter and are efficiently utilized in a range of mammalian cell lines, and obviate the need to include HMG-CoA reductase inhibitors to lower endogenous pools of mevalonate. The experimental strategy is illustrated in **Fig. 1**. A key advantage of this strategy is that F-OH and GG-OH are selectively incorporated into distinct subsets of isoprenylated proteins, providing a simple and convenient approach to label farnesylated or geranylgeranylated proteins specifically **(Fig. 2)**.

Following metabolic labeling using [^{3}H]F-OH, [^{3}H]GG-OH, or [^{3}H]mevalonolactone, the metabolically labeled proteins are exhaustively digested with pronase E to liberate the specific isoprenyl-cysteine residues **(Fig. 3)**. The identity of the isoprenylated cysteine residue can then be readily identified by normal or reverse-phase thin-layer chromatography (TLC). **Figures 4** and **5** show representative TLC analysis of isoprenyl-cysteines released from metabolically labeled cellular proteins, and from recombinant proteins that were isoprenylated in vitro.

The isoprenol labeling and pronase E difestion may also be applied to the analysis of individual metabolically labeled proteins. These methods provide a simple and convenient approach for the identification of the isoprenyl group found on a specific protein. Two experimental approaches are available. In the

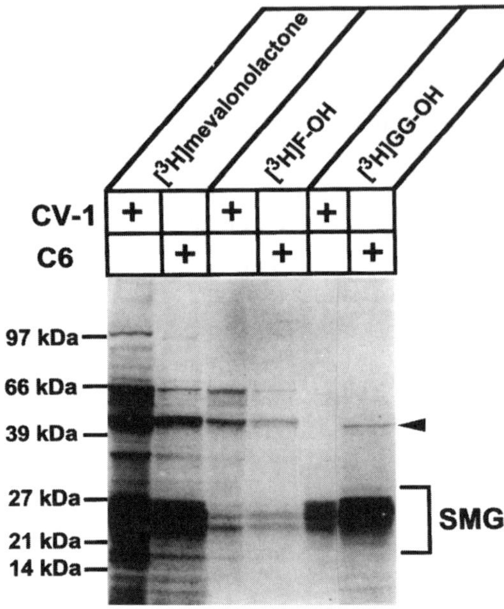

Fig. 2. SDS-PAGE analysis of proteins labeled by incubating C6 glioma cells and CV-1 cells with [^{3}H]mevalonate, [^{3}H]F-OH, or [^{3}H]GG-OH. The details of the metabolic labeling procedure and SDS-PAGE analysis are described in **Subheadings 3.1.**, **3.2.**, and **3.3.** For these analyses, proteins were metabolically labeled by incubating the indicated cultured cells with [^{3}H]F-OH in the presence of lovastatin (5 μg/mL), because it increased the amount of radioactivity incorporated into protein during long-term incubations, presumably by reducing the size of the endogenous pool of F-P-P. The gel patterns reveal that distinctly different sets of proteins are labeled by each precursor in C6 and CV-1 cells. Consistent with selective labeling by [^{3}H]F-OH and [^{3}H]GG-OH, the labeling pattern for [^{3}H]mevalonate, which can be converted to [^{3}H]-labeled F-P-P and GG-P-P, appears to be a composite of the patterns seen with the individual [^{3}H]isoprenols. SMG = proteins in the size range (19–27 kDa) of small GTP-binding proteins. Figure reprinted with permission from **ref. *14***.

first, separate cell cultures are incubated with [^{3}H]mevalonate, [^{3}H]F-OH, or [^{3}H]GG-OH, and the metabolically labeled protein of interest is then purified and subjected to SDS-PAGE analysis. The specificity of the F-OH and GG-OH incorporation allows the identity of the prenyl group to be directly assessed **(Fig. 6)**. In the second, the cells are metabolically labeled with [^{3}H]mevalonolactone, and the isolated protein of interest is subjected to pronase E treatment, followed by TLC analysis of the butanol-soluble products. In this case, analysis of the chromatogram reveals the nature of the isoprenyl group **(Fig. 7)**. The

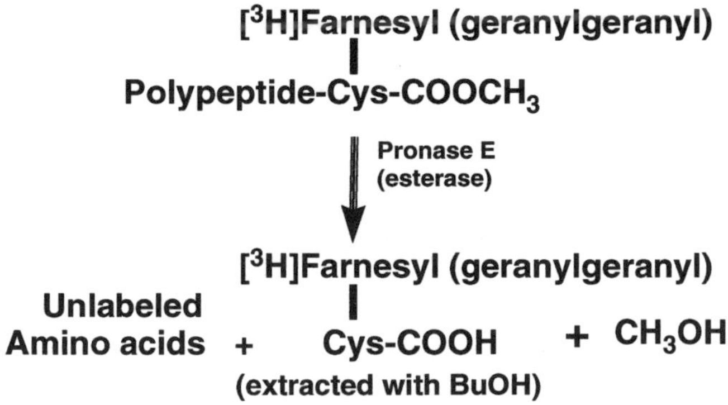

Fig. 3. Experimental scheme for the rapid analysis of metabolically labeled isoprenylated cysteine residues labeled by various isotopic precursors. The experimental details of this procedure are described in **Subheading 3.4.**

experimental methods for these analytical procedures are presented in detail in this chapter.

2. Materials

2.1. Metabolic Radiolabeling of Mammalian Cells in Culture with [^{3}H]Farnesol, [^{3}H]Geranylgeraniol, or [^{3}H]Mevalonolactone

1. ω, t, t-[1-^{3}H]Farnesol (20 Ci/mmol, American Radiolabeled Chemicals, St. Louis, MO).
2. ω, t, t, t-[^{3}H]Geranylgeraniol (60 Ci/mmol, American Radiolabeled Chemicals).
3. [^{3}H]Mevalonolactone (60 Ci/mmol, American Radiolabeled Chemicals).
4. Appropriate cell culture media, plastic ware, and cell incubator.
5. Ethanol (95%).
6. Serum Supreme (SS), an inexpensive FBS substitute obtained from BioWhittaker, has been successfully used with C6 glioma, CHO clone UT-2, and green monkey kidney (CV-1) cells in this laboratory.
7. Bath sonicator.

2.2. Delipidation of Labeled Proteins

1. Phosphate-buffered saline (PBS).
2. PBS containing 2 m*M* EDTA.
3. Methanol.
4. Chloroform–methanol (2:1, v/v).
5. Disposable conical screw-cap glass centrifuge tubes (12 mL).
6. Probe sonicator.
7. Bench-top centrifuge.

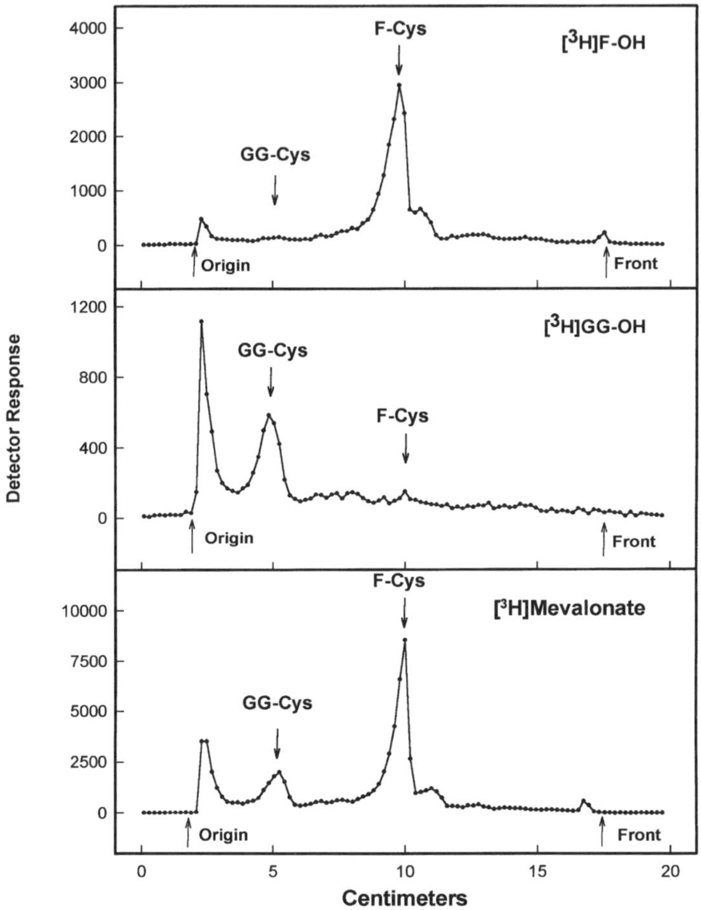

Fig. 4. Chromatographic analysis of isoprenyl-cysteine residues metabolically labeled by incubating C6 glioma cells with [^{3}H]F-OH (upper panel), [^{3}H]GG-OH (middle panel), or [^{3}H]mevalonate (lower panel). From the traces illustrated, it can be seen that F-Cys and GG-Cys are both metabolically labeled when C6 cells are incubated with [^{3}H]mevalonate, but only F-Cys or GG-Cys are labeled when C6 cells are incubated with [^{3}H]F-OH or [^{3}H]GG-OH, respectively. Virtually identical results were obtained by chromatographic analysis of the pronase E digests of CV-1 proteins metabolically labeled by each isotopic precursor. Radiolabeled products are also observed at the origin of the TLC, which could be incompletely digested isoprenylated peptides, or, in the case of [^{3}H]mevalonate and [^{3}H]GG-OH, possibly mono- or digeranylgeranylated Cys-Cys or Cys-X-Cys sequences (**ref. *11*, Fig. 5**).

2.3. SDS-PAGE Analysis of Metabolically Labeled Proteins

1. Sodium dodecyl sulfate-polyacrylamide gel electrophonesis (SDS-PAGE) apparatus.
2. Western blot transfer apparatus.
3. Nitrocellulose membrane (Schleicher and Schuell, Protran BA83).

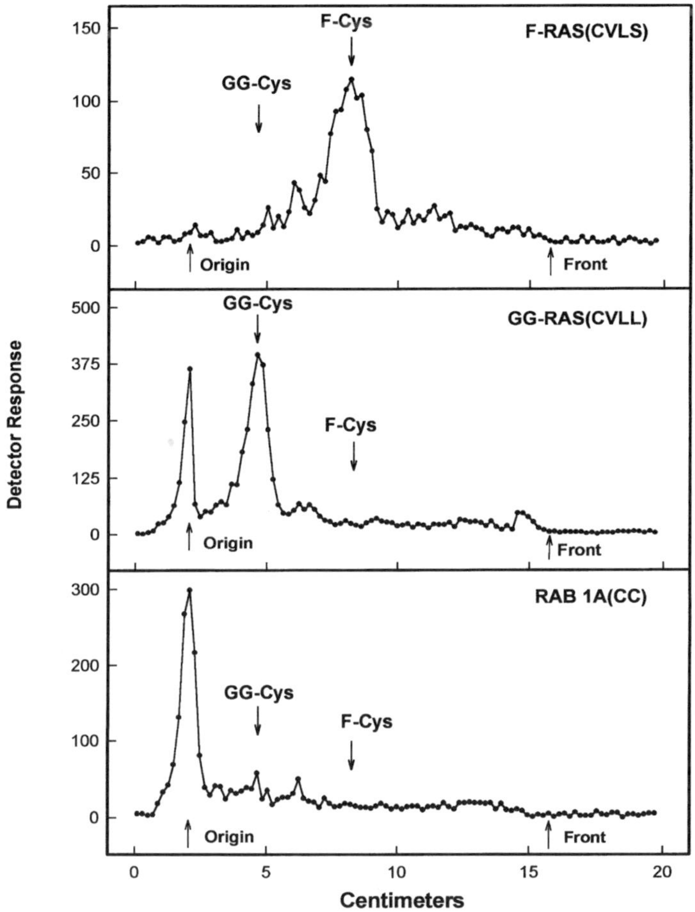

Fig. 5. Chromatographic analyses of isoprenyl-cysteine residues liberated by pronase E digestion of recombinant protein substrates enzymatically labeled in vitro by [^{3}H]F-P-P or [^{3}H]GG-P-P. The recombinant proteins were labeled by incubation with recombinant FTase (upper panel), GGTase I (middle panel), or GGTase II (lower panel), essentially under the conditions described previously ***(19)***.

4. 2% SDS, 5 m*M* 2-mercaptoethanol.
5. Ponceau S: 0.2% Ponceau S, 3% TCA, 3% sulfosalicylic acid (Sigma, S 3147).
6. Fluorographic reagent (Amplify, Amersham Corp.).
7. Sheets of stiff plastic (such as previously exposed X-ray film).

2.4. Pronase E Digestion and Chromatographic Analysis of Radiolabeled Proteins

1. HEPES, pH 7.4.
2. Calcium acetate.

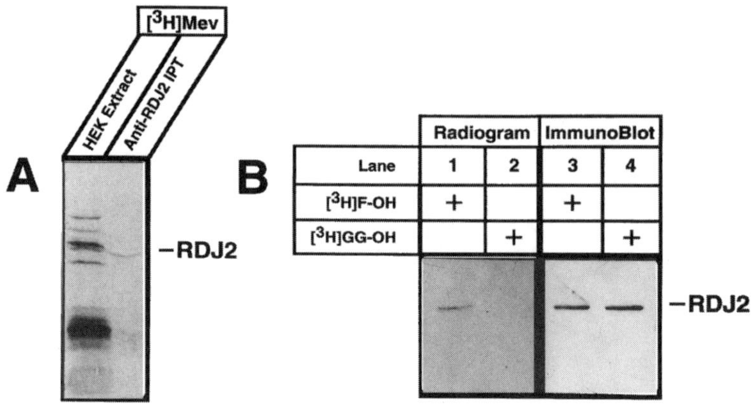

Fig. 6. Specific metabolic labeling of RDJ2 transfected HEK cells. **(A)** Monolayers of human embryonic kidney HEK cells transfected with pRDJ2 (an expression plasmid that contained the RDJ2 cDNA under the control of a CMV promoter) were radio-labeled with [^{3}H]mevalonate and the expressed RDJ2 immunoprecipitated from detergent-solubilized cell extracts, using a RDJ2-specific antibody. A portion of the resulting immunoprecipitate, as well as a portion of the cell extract, were subjected to SDS-PAGE. The gel was treated with Amplify, dried, and exposed to film for 14 d. **(B)** Immunoprecipitated RDJ2 after metabolic labeling of transfected HEK cells with [^{3}H]F-OH (lane 1) or [^{3}H]GG-OH (lane 2) was subjected to SDS-PAGE analysis and fluorography. The remaining immunoprecipitated protein fractions isolated from HEK cells after metabolic labeling with [^{3}H]F-OH (lane 3) or [^{3}H]GG-OH (lane 4) were immunoblotted, using anti-RDJ2 IgG, and subjected to chemiluminescence detection.

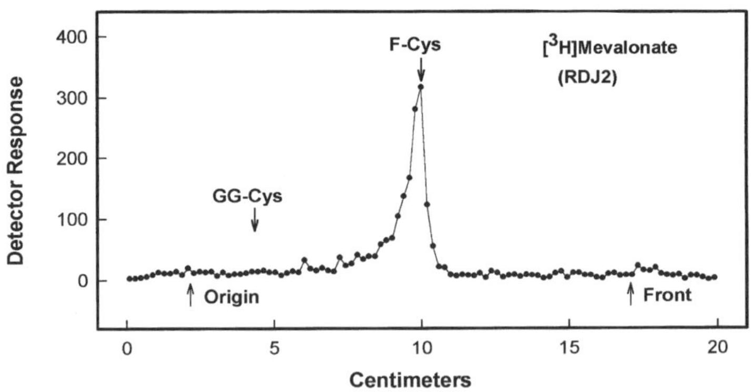

Fig. 7. Chromatographic analysis of butanol-soluble products released by pronase E digestion of RDJ2 protein metabolically labeled by incubation with [^{3}H]mevalonolactone. Immunoprecipitated RDJ2, isolated from [^{3}H]mevalonolactone-labeled HEK cells, was subjected to pronase E digestion. The labeled products were extracted with 1-butanol, and analyzed by reverse-phase chromatography using C18 reverse-phase TLC plates, and developed in acetonitrile-water-acetic acid (75:25:1). Radioactive zones were located with a Bioscan Imaging System 200-IBM. The arrows indicate the position of authentic F-Cys and GG-Cys.

3. Bath sonicator.
4. Pronase E (Sigma).
5. 37°C water bath.
6. *n*-Butanol saturated with water.
7. Bench centrifuge.
8. Oxygen-free nitrogen gas.
9. Chloroform–methanol–H_2O (10:10:3, v/v/v).
10. Farnesyl-cysteine (F-Cys) was synthesized as described by Kamiya et al. ***(24)***.
11. Geranylgeranyl-cysteine (GG-Cys) was synthesized as described by Kamiya et al. ***(24)*** except that isopropanol is substituted for methanol in the reaction solvent, to improve the yield of the synthetic reaction.
12. Silica Gel G 60 TLC plates (Sigma).
13. Chloroform–methanol–7 N ammonia hydroxide (45:50:5, v/v/v).
14. Si*C_{18} reverse-phase plates (J. T. Baker, Phillipsburg, NJ).
15. Acetonitrile–H_2O–acetic acid (75:25:1, v/v/v).
16. Conical glass tubes.
17. Anisaldehyde spray reagent ***(25)*** (*see* **Note 1**).
18. Ninhydrin spray reagent (*see* **Note 2**).

2.5. Immunoprecipitation of Specific Radiolabeled Protein

1. Cell lysis buffer: 20 m*M* Tris-HCl, pH 7.5, 150 m*M* NaCl, 1% NP-40, 1 m*M* PMSF (add fresh PMSF solution to buffer, and lysis is carried out at 4°C).
2. Protein-specific immunoprecipitating antibody.
3. Protein A Sepharose (Pharmacia, Uppsala, Sweden).
4. Wash buffer A: 20 m*M* Tris-HCl, pH 7.5, 150 m*M* NaCl, 0.2% NP-40.
5. Wash buffer B: 20 m*M* Tris-HCl, pH 7.5, 500 m*M* NaCl, 0.2% NP-40.
6. Wash buffer C: 20 m*M* Tris-HCl, pH 7.5, 150 m*M* NaCl.
7. *n*-Butanol saturated with water.
8. Table-top ultracentrifuge (Beckman, Palo Alto, CA).

3. Methods

3.1. Procedure for Selective Metabolic Labeling of Farnesylated and Geranylgeranylated Proteins

Procedures are described for the metabolic labeling of mammalian cells grown to near confluence in Falcon 3001 tissue-culture dishes. These protocols can be scaled up or down, as appropriate. The incorporation of [^{3}H]F-OH or [^{3}H]GG-OH into protein was linear with respect to time, and the concentration of [^{3}H]isoprenol in tissue-culture dishes ranging from 10 to 35 mm in diameter under the conditions described here ***(13,14)*** (*see* **Note 3**).

1. For metabolic labeling experiments with either [^{3}H]F-OH or [^{3}H]GG-OH, a disposable conical glass centrifuge tube (screw-cap) and teflon-lined cap are flame-sterilized. The labeled isoprenol dissolved in ethanol is added to the tube, and the ethanol is evaporated under a sterile stream of air.

2. An appropriate volume of sterile SS is added, to yield a final concentration of 0.5–1 mCi/mL. The labeled isoprenols are dispersed in the SS by sonication in a Branson bath sonicator for 10 min. After sonication, an aliquot is taken for liquid scintillation counting, to verify that the [^{3}H]-labeled isoprenol has been quantitatively dispersed in SS.
3. The growth medium is removed from the cultured cells by aspiration and 500 µL of labeling medium, consisting of an appropriate medium and SS (final concentration = 3–5%) containing [^{3}H]GG-OH or [^{3}H]F-OH, is added.
4. Cell cultures are typically incubated at 37°C under 5% CO_2 for 6–24 h. Actual culture media and incubation conditions will vary, depending on the specific cell type being studied.

3.2. Recovery of Metabolically Labeled Proteins from Adherent Cell Lines

1. The labeling medium is removed by gentle aspiration.
2. The metabolically labeled cells are washed with 1–2 mL ice-cold PBS, to remove unincorporated isotopic precursor.
3. One mL of PBS containing 2 m*M* EDTA is added, and cells are incubated for 5 min at room temperature. The washed cells are gently scraped from the dish with a disposable cell scraper, and transferred to a 12-mL conical glass centrifuge tube.
4. The metabolically labeled cells are sedimented by centrifugation (500*g*, 5 min), and the PBS/EDTA is removed by aspiration (avoid disrupting the cell pellet).
5. The cells are resuspended in PBS (1–2 mL), and the PBS is removed by aspiration after sedimenting the cells by centrifugation.
6. CH_3OH (2 mL) is added to the cell pellet, and the pellet is disrupted by sonication, using a probe sonicator.
7. The suspension is sedimented by centrifugation (1500*g* for 5 min).
8. The CH_3OH extract is carefully removed, to avoid disturbing the partially delipidated pellet, and transferred to a glass conical tube (*see* **Note 4**).
9. The protein pellet is re-extracted twice with 2 mL of $CHCl_3/CH_3OH$ (2:1), and the extracts are pooled with the CH_3OH extract (*see* **Note 4**).
10. Residual organic solvent is removed from the delipidated protein pellet by evaporation under a stream of nitrogen. The delipidated protein fractions are then subjected to pronase E digestion for analysis of isoprenyl-cysteine analysis by TLC (*see* **Subheading 3.4.**), or dissolved in 2% SDS, 5 m*M* 2-mercaptoethanol for SDS-PAGE analysis.

3.3. SDS-PAGE Analysis of Metabolically Labeled Proteins

To examine the molecular size and number of proteins metabolically labeled by incubation with [^{3}H]F-OH or [^{3}H]GG-OH, the delipidated protein fractions can be analyzed by SDS-PAGE. Because these experiments rely on the detection of low-energy ^{3}H-labeled compounds, two procedures are described for the use of fluorography to increase the sensitivity of detection (*see* **Note 5** before proceeding).

1. The delipidated protein fractions are solubilized in 2% SDS, 5 m*M* β-mercaptoethanol. An aliquot is used to determine the amount of labeled precursor incorporated into protein.
2. The radiolabeled polypeptides (20–60 μg protein) are analyzed by SDS-PAGE, using an appropriate percentage polyacrylamide resolving gel (4–20%) for the proteins of interest.
3. Following SDS-PAGE, gels can be analyzed using two distinct methods. In the first, the gel is directly soaked in the fluorographic reagent, Amplify (Amersham), according to the manufacturer's protocol, dried, and exposed to X-ray film, as described below **(step 5)**. In a second approach, proteins are electrophoretically transferred to nitrocellulose filters, and stained with Ponceau S to determine the efficiency of transfer. The nitrocellulose filters are then destained by brief washing in distilled water and allowed to air dry. (Please *see* **Note 5** for a discussion of the merits of each method, before continuing.)
4. The filters are then dipped briefly in the fluorographic reagent, Amplify (Amersham), placed on a sheet of plastic backing, and dried for 1 h at 50°C. It is important that a thin and even film of Amplify reagent remain on the filter, and that it be placed protein-side up to dry.
5. Fluorograms are produced by exposing preflashed X-ray film to the nitrocellulose filter, or to dried SDS-PAGE gel, for 5–30 d at –80°C.

3.4. Methods for Identification of Cysteine-Linked Isoprenyl Group

These simple methods are inexpensive, rapid, and allow the identification of the isoprenyl-cysteine residue(s) the protein(s). Examples of this method for the identification of isoprenyl-cysteine groups from metabolically labeled cells and recombinant proteins labeled in vitro are shown in **Figs. 4** and **5**. As expected, [^{3}H]F-Cys and [^{3}H]GG-Cys were liberated from RAS(CVLS) and RAS(CVLL), respectively. A radioactive peak is also seen at the origin in the analysis of the pronase digest of radiolabeled RAS(CVLL) (**Fig. 5**, middle panel). This radiolabeled product(s) is probably incompletely digested [^{3}H]geranylgeranylated peptides. Rab 1A terminates in two cysteine residues, both of which are isoprenylated ***(11)***. **Figure 5** (lower panel) indicates that pronase E is incapable of cleaving between the two cysteine residues.

1. To liberate the labeled isoprenyl-cysteine residues for analysis, the delipidated protein fractions (50–100 μg) are incubated with 2 mg of pronase E, 50 m*M* HEPES (pH 7.4)-2 m*M* calcium acetate in a total volume of 0.1 mL at 37°C for approx 16 h. The experimental scheme for this analysis is illustrated in Fig. 3 (*see* **Note 6**).
2. Proteolysis is terminated by the addition of 1 mL of n-butanol saturated with H_2O, and mixing vigorously.
3. Centrifuge the mixture for 5 min in a benchtop centrifuge at 1500*g*. Two phases will form, and the upper phase should be clear (*see* **Note 7**).

4. Transfer the upper phase, containing *n*-butanol, to a separate conical glass centrifuge tube.
5. Add 1 mL of H_2O to the butanol extract, and mix vigorously. Centrifuge at 1500*g* for 5 min, to effect a phase separation. Remove the lower aqueous phase with a Pasteur pipet (*see* **Note 8**). Evaporate the *n*-butanol under a stream of nitrogen at 30–40°C (*see* **Note 9**).
6. The labeled isoprenyl-cysteines are dissolved in 250 µL $CHCl_3/CH_3OH/H_2O$ (10:10:3, v/v/v) by mixing vigorously, and an aliquot (10 µL) is taken to determine the amount of radioactivity.
7. The radiolabeled products are analyzed chromatographically on a normal-phase system, using silica gel G 60 TLC plates developed in $CHCl_3/CH_3OH$/7 *N* NH_4OH (45:50:5, v/v/v), or by reverse-phase chromatography using silica gel Si*C18 reverse-phase plates developed with acetonitrile/H_2O/acetic acid (75:25:1,v/v/v) (*see* **Note 10**).
8. The desired developing solvent mixture is added to the chromatography tank to a depth of 0.5–1.0 cm and allowed to equilibrate.
9. Dry entire sample under nitrogen stream. Redissolve in chloroform–methanol–water (10:10:3, v/v/v).
10. Apply an aliquot (approx 10,000 dpm containing 10–12 µg authentic F-Cys and GG-Cys) to the origin on the TLC plate, using a fine glass capillary or a Hamilton syringe. The addition of the unlabeled standards improves the resolution of the metabolically labeled isoprenylated cysteines.
11. Allow the sample to dry (this can be facilitated with a stream of warm air). Complete application of the labeled isoprenyl-cysteine extract to the plate in 5 µL aliquots, allowing time for the spot to dry between applications.
12. Place the plate(s) in the pre-equilibrated chromatography tank, and after the solvent has reached the top of each plate (1–2 h), remove and allow to air-dry in a fume hood.
13. When the plates are thoroughly dried, the radioactive zones are located with a Bioscan Imaging Scanner System 200-IBM or autoradiography.
14. Standard compounds are localized by exposure of the plate to the anisaldehyde spray reagent ***(25)*** or a ninhydrin spray reagent (*see* **Note 12**).

3.5. Application of These Methods to Analysis of Individual Proteins

To illustrate the utility of this approach for individual isoprenylated proteins, these methods are applied to the analysis of a recently isolated farnesylated protein, RDJ2 (rat DnaJ homolog 2). The cDNA clone of this DnaJ-related protein was recently identified ***(26)***. The predicted amino acid sequence is found to terminate with the tetrapeptide Cys-Ala-His-Gln, which conforms to the consensus sequence for recognition by protein farnesyltransferase, and was shown to undergo farnesylation in vivo.

In order to perform this analysis, a means of specifically identifying the protein of interest must be available. In this example, a protein-specific immunoprecipitating antibody was used to isolate the protein from metabolically labeled mammalian cells. However, other experimental approaches are available (*see* **Note 13** for discussion of these strategies).

1. Mammalian cells are grown to near confluence, and metabolically labeled with either [^{3}H]mevalonate, [^{3}H]F-OH, or [^{3}H]GG-OH, as described in **Subheading 3.1.** If a recombinant protein is to be analyzed, the mammalian cells should be transfected either stably or transiently with the expression vector, prior to labeling *(27)*.
2. The metabolically labeled cells are washed with 1–2 mL of ice-cold PBS, to remove unincorporated isotopic precursor.
3. One mL of PBS containing 2 m*M* EDTA is added, and incubated for 5 min at room temperature. The washed cells are gently scraped from the dish with a disposable cell scraper, and transferred to a 12-mL conical glass centrifuge tube.
4. The metabolically labeled cells are sedimented by centrifugation (500*g*, 5 min), and the PBS/EDTA is removed by aspiration (avoid disrupting the cell pellet).
5. The cells are resuspended in lysis buffer (2 mL), disrupted by passage through a 20-gage needle (3–4X), and centrifuged for 15 min at 100,000*g* in a Beckman table-top ultracentrifuge.
6. For immunoprecipitation of recombinant RDJ2, 20 µg of rabbit anti-RDJ2 antibody was added to the cleared supernatant, and incubated for 12 h at 4°C with gentle rocking (*see* **Note 14**).
7. Immune complexes were then precipitated by addition of 100 µL of a 50% slurry of protein A Sepharose for 2 h at 4°C with gentle rocking.
8. Protein A beads were collected by centrifugation (1 min in tabletop microcentrifuge at 1000*g*). The pellet was washed 3× by resuspending in 1 mL wash buffer A, 1 mL wash buffer B, and 1 mL wash buffer C, sequentially.
9. The protein is dissolved in 2% SDS, 5 m*M* 2-mercaptoethanol for SDS-PAGE analysis (*see* **Subheading 3.3.** and **Fig. 6**). Alternatively, the protein A beads are subjected to pronase E digestion for analysis of isoprenyl-cysteine analysis by TLC (*see* **Subheading 3.4.**, **Fig. 7**, and **Note 15**).

4. Notes

1. Anisaldehyde spray reagent contains 10 mL anisaldehyde, 180 mL 95% ethanol, and 10 mL concentrated sulfuric acid, added in that order.
2. Ninhydrin spray reagent contains 0.2% ninhydrin in 95% ethanol.
3. All mammalian cells tested (CHO, C6 glioma cells, and green monkey kidney (CV-1) cells) utilized F-OH and GG-OH for protein isoprenylation, except murine B-cells before or after activation by LPS. Thus, it is possible that the salvage pathway for F-OH and GG-OH (Fig. 1) may not be ubiquitous in mammalian cells.
4. The pooled organic extracts can be used for analysis of lipid products metabolically labeled by [^{3}H]F-OH (*see* **ref.** ***14***). The organic solvent is evaporated under a stream of nitrogen, and the lipid residue redissolved in $CHCl_3$–CH_3OH (2:1), containing 20 µg each of authentic cholesterol and squalene. An aliquot is taken

to determine the amount of radioactivity incorporated into the lipid extracts. The lipid products are analyzed on Merck silica gel G 60 TLC plates (Sigma) by developing with hexane–diethyl ether–acetic acid (65:35:1) or chloroform. Radioactive zones were located with a Bioscan Imaging Scanner System 200-IBM. Standard compounds are located with iodine vapor or anisaldehyde spray reagent ***(25)***.

5. Western transfer is preferred, because transferring labeled proteins to nitrocellulose membrane appears to give a gain of 2–10-fold in sensitivity. One suspects that the polyacrylamide gel acts to quench the signal from radiolabeled protein. The transfer step serves to collect proteins in a single plane, and eliminates this problem. However, care should be taken with the interpretation of these experiments. It is possible that some radiolabeled proteins, particularly those of either small (<10 kDa) or very large (>100 kDa) mol, may be inefficiently transferred. The properties of the protein, percentage of acrylamide, transfer buffer components, and transfer time will each influence the transfer efficiency. Direct analysis of the gel will be less sensitive, requiring more labeled protein and longer exposure times, but will not suffer from losses of material during transfer.
6. Pronase E will completely dissolve over a period of 30 min at 37°C, and these protease preparations contain sufficient esterase activity to hydrolyze the carboxymethyl esters at the C-termini of isoprenylated proteins ***(28)***.
7. The lower aqueous phase remains cloudy, and contains precipitated proteins and peptides.
8. The addition of H_2O will significantly reduce the volume of the *n*-butanol phase at this step (reducing the time required for the following steps).
9. The addition of an equal volume of *n*-hexane to the *n*-butanol will speed this evaporation, by forming an azeotrope. The addition of the hexane will cause the solution to become cloudy and biphasic. The upper (butanol/hexane phase) evaporates quite rapidly, and the subsequent addition of 1 mL of 100% ethanol speeds the evaporation of the lower aqueous phase.
10. The resolution of GG-Cys and F-Cys is better in the reverse phase chromatography system. If the normal-phase system is used, the plates should be activated by heating in a 100°C oven for at least 1 h.
11. At least 5000 dpm are required for good detection, using a Bioscan Imaging System, but analysis will be better (and faster) with 10,000 dpm or more. The entire sample from the digestion of 100 µg of labeled protein can usually be loaded on the TLC plates without any significant loss of chromatographic resolution.
12. Spray until plate is moist, and heat in 100°C oven until spots appear. The plates are scanned prior to spraying to avoid the reagent vapors, and because the spray reagents quench the detection of radioactivity.
13. A consideration before beginning these experiments is the abundance of the protein of interest. The analysis of a very low abundance protein will require a large number of radiolabeled tissue-culture cells. Therefore, the overproduction of the protein using a mammalian expression vector may present a distinct

experimental advantage. This approach also allows the cDNA to be modified to contain a unique epitope or affinity sequence at the N-terminus. In this way, proteins for which specific antibodies are not available may be studied.

14. Optimal immunoprecipitation conditions must be established for each protein–antibody complex.
15. Pronase E digestion can be carried out directly on the protein A beads without further processing. Follow directions given in **Subheading 3.4.**, scaling up the volume of the proteolysis reaction mixture to provide sufficient liquid to amply cover the protein A beads. Follow **steps 2–15** as directed. Initially, proteins should be labeled with mevalonolactone and both free isoprenols, to ensure correct interpretation of the results. Although a variety of proteins have been tested (*see* **Figs. 2**, **4**, and **7**) using [^{3}H]isoprenol labeling, the list of individual proteins is quite limited. It will be necessary to analyze a wide variety of defined isoprenylated proteins to further establish the reliability and limitations of this method.

Acknowledgments

The methods described in this chapter were developed with support from NIH grants GM36065 (C.J.W) and EY11231 (D.A.A).

References

1. Maltese, W. A. (1990) Posttranslational modification of proteins by isoprenoids in mammalian cells. *FASEB J.* **4,** 3319–3328.
2. Glomset, J. A., Gelb, M. H., and Farnsworth, C. C. (1990) Prenyl proteins in eukaryotic cells: a new type of membrane anchor. *TIBS* **15**, 139–142.
3. Clarke, S. (1992) Protein isoprenylation and methylation at carboxyl-terminal cysteine residues. *Annu. Rev. Biochem.* **61,** 355–386.
4. Schafer, W. R. and Rine, J. (1992) Protein prenylation: genes, enzymes, targets, and functions. *Annu. Rev. Genet.* **30,** 209–237.
5. Zhang, F. L. and Casey, P. J. (1996) Protein prenylation: molecular mechanisms and functional consequences. *Annu. Rev. Biochem.* **65,** 241–269.
6. Hancock, J. F., Magee, A. I., Childs, J. E., and Marshall, C. (1989) All ras proteins are polyisoprenylated but only some are palmitoylated. *Cell* **57,** 1167–1177.
7. Hancock, J. F., Paterson, H., and Marshall, C. J. (1990) A polybasic domain or palmitoylation is required in addition to the CAAX motif to localize $p21^{ras}$ to the plasma membrane. *Cell* **63,** 133–139.
8. Hancock, J. F., Cadwallader, K., and Marshall, C. J. (1991) Methylation and proteolysis are essential for efficient membrane binding of prenylated $p21^{K\text{-}ras(B)}$. *EMBO J.* **10,** 641–646.
9. Schafer, W. R., Kim, R., Sterne, R., Thorner, J., Kim, S-H., and Rine, J. (1989) Genetic and pharmacological suppression of oncogenic mutations in RAS genes of yeast and humans. *Science* **245,** 379–385.
10. Kato, K., Cox, A. D., Hisaka, M. M., Graham, S. M., Buss, J. E., and Der, C. J. (1992) Isoprenoid addition to Ras protein is the critical modification for its

membrane association and transforming activity. *Proc. Natl. Acad. Sci. USA* **89,** 6403–6407.

11. Seabra, M. C., Goldstein, J. L., Sudhof, and Brown, M. S. (1992) Rab geranylgeranyl transferase: a multisubunit enzyme that prenylates GTP-binding proteins terminating in cys-x-cys or cys-cys. *J. Biol. Chem.* **267,** 14,497–14,503.
12. Grunler J., Ericsson, J., and Dallner, G. (1994) Branch-point reactions in the biosynthesis of cholesterol, dolichol, ubiquinone and prenylated proteins. *Biochim. Biophys. Acta* **1212,** 259–277.
13. Crick, D. C., Waechter, C. J., and Andres, D. A. (1994) Utilization of geranylgeraniol for protein isoprenylation in C6 glial cells. *Biochem. Biophys. Res. Commun.* **205,** 955–961.
14. Crick, D. C., Andres, D. A., and Waechter, C. J. (1995) Farnesol is utilized for protein isoprenylation and the biosynthesis of cholesterol in mammalian cells. *Biochem. Biophys. Res. Commun.* **211,** 590–599.
15. Baxter, A., Fitzgerald, B. J., Hutson, J. L., McCarthy, A. D., Motteram, J. M., Ross, B. C., et al. (1992) Squalestatin 1, a potent inhibitor of squalene synthase, which lowers serum cholesterol *in vivo*. *J. Biol. Chem.* **267,** 11,705–11,708.
16. Bergstrom, J. D., Kurtz, M. M., Rew, D. J., Amend, A. M., Karkas, J. D., Bostedor, R. G., et al. (1993) Zaragozic acids: a family of fungal metabolites that are picomolar competitive inhibitors of squalene synthase. *Proc. Natl. Acad. Sci. USA* **90,** 80–84.
17. Hasumi, K., Tachikawa, K., Sakai, K., Murakawa, S., Yoshikawa, N., Kumizawa, S., and Endo, A. (1993) Competitive inhibition of squalene synthetase by squalestatin 1. *J. Antibiot.* (Tokyo) **46,** 689–691.
18. Thelin, A., Peterson, E., Hutson, J. L., McCarthy, A. D., Ericcson, J., and Dallner, G. (1994) Effect of squalestatin 1 on the biosynthesis of the mevalonate pathway lipids. *Biochim. Biophys. Acta* **1215,** 245–249.
19. Crick, D. A., Suders, J., Kluthe, C. M., Andres, D. A., and Waechter, C. J. (1995) Selective inhibition of cholesterol biosynthesis in brain cells by squalestatin 1. *J. Neurochem.* **65,** 1365–1373.
20. Inoue, H., Korenaga, T., Sagami, H., Koyama, T., and Ogura, K. (1994) Phosphorylation of farnesol by a cell-free system from *Botryococcus brauni*. *Biochem. Biophys. Res. Commun.* **200,** 1036–1041.
21. Ohnuma, S-I., Watanabe, M., and Nishino, T. (1996) Identification and characterization of geranylgeraniol kinase and geranylgeranyl phosphate kinase from the archebacterium *Sulfolobus acidocaldarius*. *J. Biochem.* (*Tokyo*) **119,** 541–547.
22. Westfall, D., Aboushadi, N., Shackelford, J. E., and Krisans, S. K. (1997) Metabolism of farnesol: phosphorylation of farnesol by rat liver microsomal and peroxisomal fractions. *Biochem. Biophys. Res. Commun.* **230,** 562–568.
23. Crick, D. C., Andres, D. A., and Waechter, C. J. (1997) Novel salvage pathway utilizing farnesol and geranylgeraniol for protein isoprenylation. *Biochem. Biophys. Res. Commun.* **237,** 483–487.

24. Kamiya, Y., Sakurai, A., Tamura, S., Takahashi, N., Tsuchiya, E., Abe, K., and Fukui, S. (1979) Structure of rhodotorucine A, a peptidyl factor, inducing mating tube formation in *Rhodosporidium toruloides. Agric. Biol. Chem.* **43,** 363–369.
25. Dunphy, P. J., Kerr, J. D., Pennock, J. F., Whittle, K. J., and Feeney, J. (1967) The plurality of long chain isoprenoid (polyprenols) alcohols from natural sources. *Biochim. Biophys. Acta* **13,** 136–147.
26. Andres, D. A., Shao, H., Crick, D. C., and Finlin, B. S. (1997) Expression cloning of a novel farnesylated protein, RDJ2, encoding a DnaJ protein homologue. *Arch. Biochem. Biophys.* **346,** 113–124.
27. Sambrook, J., Fritsch, E. F., and Maniatis, T. (1989) *Molecular Cloning: A Laboratory Manual.* Cold Spring Harbor Laboratory, Cold Spring Harbor, NY.
28. Stimmel, J. B., Deschenes, R. J., Volker, C., Stock, J., and Clarke, S. (1990) Evidence for an S-farnesylcysteine methyl ester at the carboxyl terminus of the *Saccharomyces cerevisiae* RAS2 protein. *Biochemistry* **29,** 9651–9659.
29. Bansal, V. S. and Vaidya, S. (1994) Characterization of two distinct allyl pyrophosphatase activities from rat liver microsomes. *Arch. Biochem. Biophys.* **315,** 393–399.

10

Incorporation of Radiolabeled Prenyl Alcohols and Their Analogs into Mammalian Cell Proteins

A Useful Tool for Studying Protein Prenylation

Alberto Corsini, Christopher C. Farnsworth, Paul McGeady, Michael H. Gelb, and John A. Glomset

1. Introduction

Prenylated proteins comprise a diverse family of proteins that are post-translationally modified by either a farnesyl group or one or more geranyl-geranyl groups *(1–3)*. Recent studies suggest that members of this family are involved in a number of cellular processes, including cell signaling *(4–6)*, differentiation *(7–9)*, proliferation *(10–12)*, cytoskeletal dynamics *(13–15)*, and endocytic and exocytic transport *(4,16,17)*. The authors' studies have focused on the role of prenylated proteins in the cell cycle *(18)*. Exposure of cultured cells to competitive inhibitors (statins) of 3-hydroxy-3-methylglutaryl Coenzyme A (HMG-CoA) reductase not only blocks the biosynthesis of mevalonic acid (MVA), the biosynthetic precursor of both farnesyl and geranylgeranyl groups, but pleiotropically inhibits DNA replication and cell-cycle progression *(10,18–20)*. Both phenomena can be prevented by the addition of exogenous MVA *(10,18,19)*. The authors have observed that all-*trans*-geranylgeraniol (GGOH) and, in a few cases, all-*trans*-farnesol (FOH) can prevent the statin-induced inhibition of DNA synthesis *(21)*. In an effort to understand the biochemical basis of these effects, the authors have developed methods for the labeling and two-dimensional gel analysis of prenylated proteins that should be widely applicable.

Because of the relative diversity of prenylated proteins, it is important to use analytical methods that differentiate between them. A useful approach discussed here, and in Chapter 2 of in this volume, is to selectively label

From: *Methods in Molecular Biology, Vol. 116: Protein Lipidation Protocols*
Edited by: M. H. Gelb © Humana Press Inc., Totowa, NJ

farnesylated or geranylgeranylated proteins using [^{3}H] labeled FOH or GGOH *(22–24)*, followed by one-dimensional SDS-PAGE or high-resolution two-dimensional gel electrophoresis (2DE) of the labeled proteins. Since the enzymes that transfer prenyl groups to proteins utilize the corresponding prenyl alcohol pyrophosphate (FPP or GGPP) as substrate, these prenols are thought to undergo two phosphorylation steps prior to their subsequent utilization *(2,3)*. The discovery of a GGOH kinase and a geranylgeranyl phosphate kinase in eubacteria *(25)*, together with the ability of rat liver microsomal and peroxisomal fractions to form FPP *(26)*, provide additional evidence that these prenol pools serve as a source of lipid precursor for protein prenylation. When proteins are labeled in this way, and are subsequently analyzed by one-dimensional SDS-PAGE, it is possible to distinguish several major bands of radioactivity that correspond to two apparently distinct subsets of proteins: those that incorporate [^{3}H]-FOH and those that incorporate [^{3}H]-GGOH. But when the labeled proteins are analyzed by high-resolution 2DE, the number of radioactive proteins that are observed is at least five-fold greater, and three subsets of prenylated proteins can be identified: one subset of proteins that incorporates only farnesol, a second that incorporates only geranylgeraniol, and a third that can incorporate either prenol.

In this chapter, the advantages of using labeled prenols to dissect the differential effects of FOH and GGOH on cellular function will be presented in the context of studies of the role of prenylated proteins in cell-cycle progression. The ability of mammalian cells to incorporate natural and synthetic prenol analogs (**Fig. 1**) into specific proteins also will be discussed. In **Subheading 4.** some of the advantages and limitations of the methods will be discussed.

2. Materials

2.1. Prenols and Analogs: Synthesis and Labeling

1. All-*trans*-FOH, d^{20} 0.89 g/mL; mevalonic acid lactone; geraniol, d^{20} 0.89 g/mL (Sigma, St. Louis, MO).
2. Tetrahydrofarnesol was a gift from Hoffman la Roche (Basel, Switzerland). A racemic form can be made from farnesyl acetone following the procedure for hexahydro-GGOH (*see* **Subheading 3.1.1.**).
3. All-*trans*-GGOH, d^{20} 0.89 g/mL; all-*trans*-GGOH [1-^{3}H], 50–60 Ci/mmol; all-*trans*-FOH [1-^{3}H], 15–20 Ci/mmol; geraniol [1-^{3}H], 15–20 Ci/mmol (American Radiolabeled Chemicals, St. Louis, MO).
4. Mevalonolactone RS-[5-^{3}H(N)], 35.00 Ci/mmol (NEN, Boston, MA).
5. Charcoal; $LiAlH_4$ powder; MnO_2; $NaBH_4$ powder; triethyl phosphonoacetate; phosphonoacetone; sodium ethoxide (Aldrich, Milwaukee, WI).
6. [^{3}H]-$NaBH_4$ solid, 70 Ci/mmol (Amersham, Arlington Heights, IL).
7. Silica gel 60 (F254) (Merck, Gibbstown, NJ)
8. Aquamix liquid scintillation solution (ICN Radiochemicals, Irvine, CA).

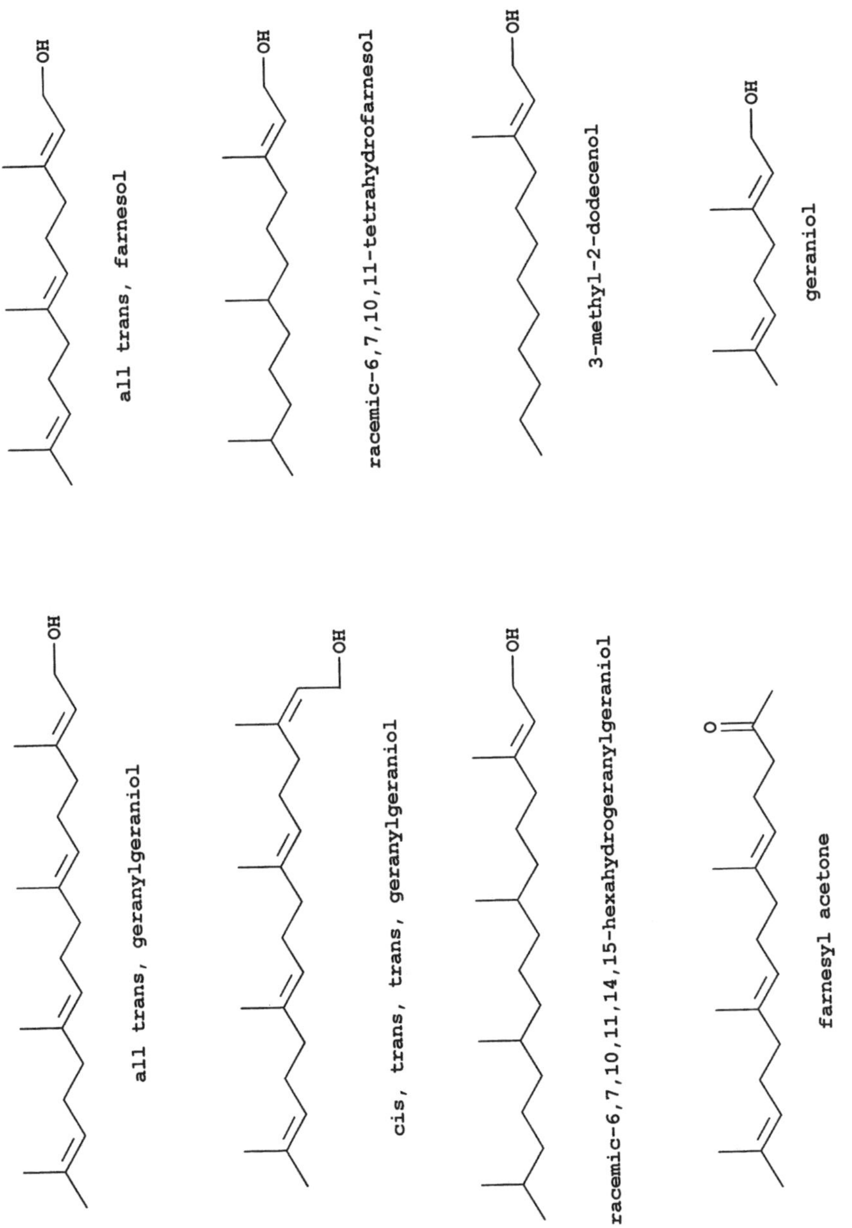

Fig. 1. Structure of the natural and synthetic prenol analogs used in these labeling studies.

2.2. Cell Culture Reagents

1. Dulbecco's modified Eagle media (DMEM) (glucose 4.5 g/L); penicillin (10,000 U/mL)–streptomycin (10 mg/mL); trypsin (0.25% w/v)–1 m*M* ethylenediaminetetraacetate (EDTA); nonessential amino acids (NEAA) solution 10 m*M*; phosphate-buffered saline (PBS) without Ca^{2+} and Mg^{2+}; fetal calf serum (FCS) (Gibco, Grand Island, NY).
2. Disposable culture Petri dishes (100 × 10 mm and 35 × 10 mm) (Corning Glass Works, Corning, NY).
3. Filters, 0.22 µm (Millipore, Bedford, MA).
4. Plasma-derived, bovine serum (PDS) (Irvine Scientific, Santa Ana, CA).
5. Aquasol scintillation cocktail; thymidine [methyl-^{3}H], 2 Ci/mmol] (NEN).
6. Trichloroacetic acid (TCA) (J. T. Baker, Phillipsburg, NJ).
7. Phenylmethylsulfonyl fluoride (PMSF), aprotinin, leupeptin, and pepstatin A (Sigma).
8. Simvastatin in its lactone form (gift from Merck, Sharp and Dohme; Rahway, NJ) is dissolved in 0.1 *M* NaOH (60°C, 3 h), to give the active form. Adjust the pH to 7.4 and the concentration to 50 m*M*, then sterilize by filtration.
9. Swiss 3T3-albino mouse cell line (3T3) were from American Type Culture Collection (ATCC), Rockville, MD.
10. Human skin fibroblasts (HSF) are grown from explants of skin biopsies obtained from healthy individuals. The cells are used between the fifth and fifteenth passages.

2.3. One-Dimensional and Two-Dimensional Gel Electrophoresis

1. SDS, TEMED, ammonium persulfate, mol wt protein standards, glycine, Bradford protein assay kit, heavyweight filter paper, and Bio Gel P6-DG (Bio-Rad, Hercules, CA).
2. Duracryl™ preblended acrylamide solution (30% T, 0.65% C, used for both one-dimensional and 2DE gels) and all other 2DE gel reagents, were obtained from ESA, Chelmsford, MA.
3. Ethylmaleimide, N-[ethyl-1,2-^{3}H] ([^{3}H]-NEM), 57 Ci/mmol (NEN).
4. N-ethylmaleimide (Aldrich).
5. Soybean trypsin inhibitor and SDS-7 mol wt standards used with 2DE gels (Sigma).
6. Amplify and Hyperfilm (Amersham).
7. Reacti-Vials™ (Pierce, Rockford, IL).
8. DNase1 and RNaseA (Worthington Biochem, Freehold, NJ).

3. Methods

3.1. Synthesis and Labeling of Prenols and Analogs

3.1.1. Synthesis of Racemic 6,7,10,11,14,15 Hexahydrogeranylgeraniol

1. This synthesis is outlined in **Fig. 2**. Weigh 10.4 g of farnesyl acetone and 500 mg 10% Pd on charcoal, dissolve in 100 mL methanol and hydrogenate at 50 psi in a

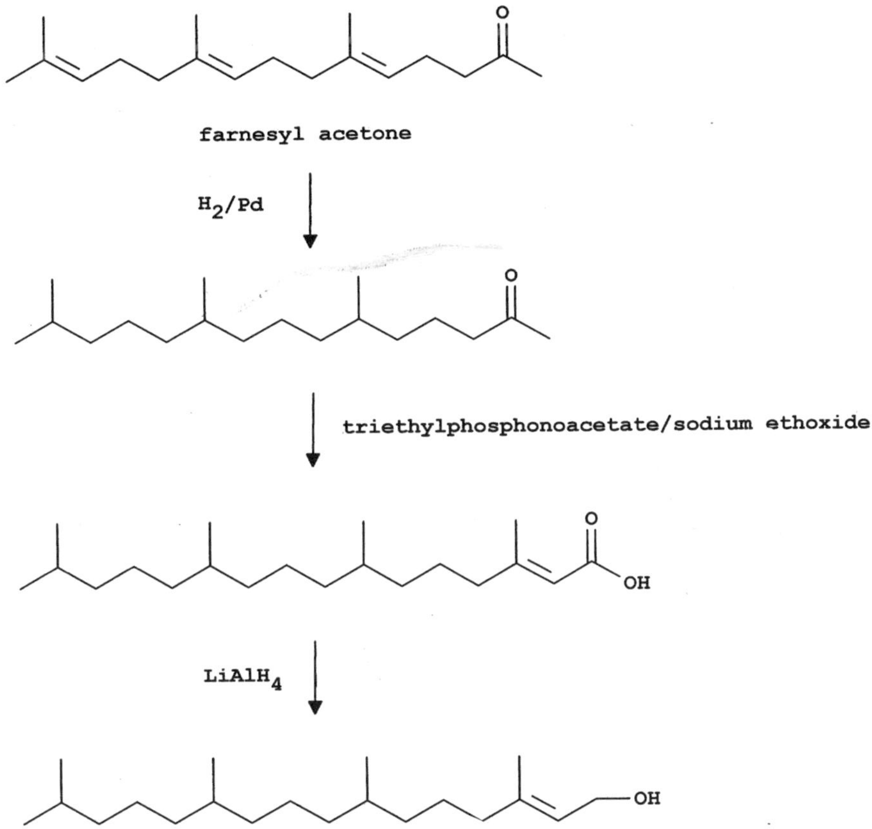

Fig. 2. Schematic for the synthesis of racemic 6,7,10,11,14,15 hexahydrogeranylgeraniol from farnesyl acetone.

Parr hydrogenator for 2 d. Analysis by TLC (7:3 hexanes/diethyl ether) and GC-MS should indicate that the material is fully converted to the hexahydrofarnesyl acetone.

2. Filter the reaction mixture over filter paper, wash with ethanol, rotary-evaporate to dryness, and take up in a small volume of ethanol, then filter over Fluorosil to remove the last traces of catalyst. Concentrate the resultant oil to dryness with a rotary evaporator (yield 8.64 g, 81% recovery).
3. Dissolve 1.56 g hexahydrofarnesyl acetone in 10 g dry dimethylformamide in a round-bottom flask submerged in ice water and equipped with a dropping funnel containing 1.37 mL triethyl phosphonoacetone. Flush the entire apparatus with argon (Ar). Add triethyl phosphonoacetone over the course of 45 min, followed by 2.2 mL 21% (w/w) sodium ethoxide in ethanol, which is added over the course of 1 h.

4. Remove the mixture from the ice water bath and continue to stir for 48 h under Ar. Analysis by TLC (7:3 hexanes/diethyl ether) should show that the majority of the starting material (R_f 0.5) has reacted (final product R_f 0.6).
5. Transfer the reaction mixture to a separatory funnel with hexane, and wash 2× with a NaCl-saturated solution. Dry the organic layer with $MgS0_4$, pass through filter paper, and evaporate the solvent to dryness by rotary evaporation (yield 1.41 g, 72% recovery).
6. Dissolve 0.75 g of the product from the previous step in 10 mL anhydrous diethyl ether in a round-bottom flask equipped with a stir bar. Cool the apparatus by stirring in an ice-water bath.
7. Add 267 mg of $LiAlH_4$, and stir the reaction mixture overnight under Ar. The following day, add 25 mL saturated NH_4Cl and stir the reaction mixture overnight under Ar.
8. Transfer the reaction mixture to a separatory funnel with diethyl ether and wash with water and saturated NaCl solution. Dry the organic phase with $MgSO_4$, filter through filter paper and concentrate to dryness (yield 440 mg, 51% recovery).
9. Purify the material by TLC (7:3 hexanes–diethyl ether, R_f 0.25), or use in the crude state if proceeding on to make radiolabeled material.

3.1.2. Synthesis of cis-Isomers

1. Form the *cis*-isomers of the allylic alcohols from the respective *trans*-alcohols after oxidation to the aldehyde (*see* **Subheading 3.1.3.**). Allow the mixture of *cis/trans* isomers to come to equilibrium (approx 36 *cis*:65 *trans*) by standing at room temperature for several days.
2. Purify the *cis*-isomer by TLC (7:3 hexanes/diethyl ether).

3.1.3. Tritium Labeling of Prenols

1. The method for radiolabeling the prenols is outlined in **Fig. 3**. First, oxidize the alcohol to the aldehyde with an excess of MnO_2. Dissolve 44.7 mg of hexahydrogeranylgeraniol in 2 mL benzene, and add 517 mg of MnO_2. Mix the reaction by tissue culture rotator overnight.
2. The next day, allow the reaction mixture to settle, remove the liquid phase, and filter over glass wool to remove any remaining MnO_2. Add diethyl ether to the original reaction vessel, and repeat the process. Combine the two solutions, and concentrate to dryness under a stream of nitrogen.
3. Chromatograph the material by TLC (7:3 hexanes/diethyl ether, R_f 0.35). Two overlapping bands are present corresponding to the *cis*- and *trans*-double bond isomer, the upper band being the *cis*-isomer and the lower band being the *trans*.
4. Remove the lower one-third of the overlapping bands, elute with diethyl ether, dry with $MgSO_4$, and concentrate to dryness (yield 20%). If not used immediately, this material should be stored at –70°C to prevent isomerization of the *trans*-aldehyde to the equilibrium mixture (~30% *cis*).
5. Elute the remaining material, and rechromatograph if desired.

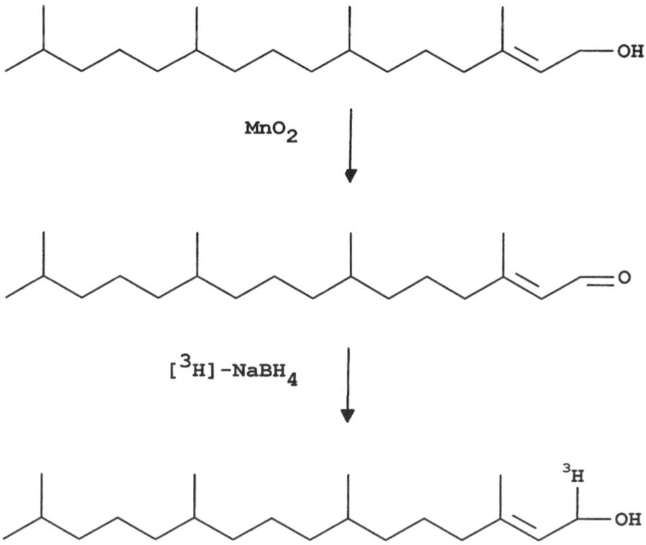

Fig. 3. Schematic for the tritium labeling of naturally occurring prenols.

6. Reduce the aldehyde to the alcohol using [^{3}H]NaBH_4. Dissolve 2 mg of the aldehyde in 1 mL absolute ethanol, and add 5 µL 14 N NH_4OH. Dissolve the [^{3}H]NaBH_4 in ethanol at a concentration of 100 mCi/mL. Add 200 µL of the [^{3}H]NaBH_4 to the aldehyde solution, shake the solution, and then leave it vented in the hood for at least 4 h.
7. Concentrate the material to dryness in a gentle stream of Ar, take up in diethyl ether, and place over a column of silica gel overlaid with $MgSO_4$. Purify the material by TLC (7:3 hexanes/diethyl ether) to remove unreacted starting material and the small amount of the *cis*-isomer that is generated during the reaction. Elute the labeled alcohols from the silica using diethyl ether. Dry using a stream of Ar.
8. Dissolve the product in ethanol, and bring to a final concentration of 2.1 m*M*, 2.6 m*M*, 6.5 m*M*, or 0.4 m*M* for labeled 2 *cis*-GGOH, hexahydrogeranylgeraniol, geraniol, and tetrahydrofarnesol, respectively.

3.2. Cell Culture Experiments

1. Grow HSF and 3T3 cells in monolayers, and maintain in 100-mm Petri dishes at 37°C in a humidified atmosphere of 95% air, 5% CO_2 in DMEM, pH 7.4, supplemented with 10% FCS v/v, 1% (v/v) NEAA, penicillin (100 U/mL), and streptomycin (0.1 mg/mL).
2. Dissociate confluent stock cultures with 0.05% trypsin-0.02% EDTA, and seed HSF cells (3 × 10^5 cells/35-mm Petri dish) or 3T3 cells (2 × 10^5 cells/35-mm Petri dish) in a medium containing 0.4% FCS or 1% PDS, respectively, to stop cell replication.

3. HSF cells become quiescent within 3 d and the experiments can begin on d 4. For 3T3 cells, change the medium on d 2 and 4; cells become quiescent within 5 d, and the experiments can begin on d 6.
4. At this time, stimulate the cells by replacing the medium with one containing 10% FCS, in the presence or absence of the tested compounds, and continue the incubation as needed at 37°C. Simvastatin is used at a final concentration of 40 μ*M*, when required.
5. Dissolve unlabeled prenols in absolute ethanol and prepare stock solutions as follows:
   a. 2 m*M* all-*trans*-GGOH (the density for this and all other prenols used in these studies is d^{20} 0.89 g/mL). Add 30 μL of the prenol to 49 mL ethanol and store in 1-mL aliquots at –20°C. On the day of the experiment, dilute an aliquot with ethanol (e.g., 300 μL of GGOH to 900 μL of ethanol) to obtain a working solution of 0.5 m*M*.
   b. 4 m*M* all-*trans*-FOH. Add 20 μL of the prenol to 20 mL of ethanol and store in 1-mL aliquots at –20°C. On the day of the experiment, dilute this 1:3 with ethanol to make 1 m*M* working stock solution.
   c. 4 m*M* geraniol. Add 20 μL of the prenol to 28.83 mL ethanol and store in 1-mL aliquots at –20°C. On the day of the experiment, dilute this 1:3 with ethanol to make 1 m*M* working stock solution.
6. Prenols are light-sensitive, so experiments should be performed under dim light, and incubation should be done with Petri dishes covered with foil. Be careful not to exceed 1% (v/v) ethanol in the culture medium.
7. For estimation of DNA synthesis after mitogenic stimulation, incubate cells for 22–24 h at 37°C then change the medium to one containing 10% FCS and 2 μCi/mL [^{3}H]-thymidine and incubate the cells for another 2 h at 37°C ***(19)***. Remove the medium and wash the monolayers once with PBS at room temperature, then add 2 mL of fresh, ice-cold 5% TCA (w/v), and keep the cells at 4°C on ice for at least 10 min. Remove TCA and wash once with 5% TCA, and dissolve the monolayers in 1 mL of 1 N NaOH for 15 min. Transfer 500 μL of the cell lysates to a liquid scintillation vial, add 150 μL glacial acetic acid and 5 mL Aquasol, and measure incorporated radioactivity in a liquid scintillation counter (**Table 1**; **Fig. 4,** *see* page 134).

3.3. Cell Labeling and Prenylated Protein Analysis

3.3.1. Labeling of Proteins with [^{3}H] Farnesol or [^{3}H] Geranylgeraniol and One-Dimensional SDS-PAGE Analysis

1. Dry 9 mCi [^{3}H]-MVA, 500 μCi [^{3}H]-GGOH, or 1 mCi [^{3}H]-FOH, and resuspend in 50 μL ethanol, to bring to the concentration of 4.2 m*M*, 166 μ*M*, or 1.1 m*M*, respectively.
2. Incubate the cells (3 × 35-mm Petri dishes/sample) (*see* **Notes 2–6**) for 20 h at 37°C, with the appropriate labeled isoprenoids. Incubation periods as short as 5 h are sufficient to label most proteins, making it possible to perform time course experiments.

Table 1
Mevalonate, Prenols, and Synthetic Analogs Vary in Their Ability to Prevent the Inhibitory Effect of Simvastatin on DNA Synthesis in Cultured Cells as Measured by Incorporation of [^{3}H]-Thymidine

Cell type	HSF	3T3
Treatment Conditions:	(Reported as % control)	
Simvastatin (S) 40 μ*M*	30	30
S + mevalonate 100 μ*M*	100	90
S+ all-*trans*-GGOH 5 μ*M*	85	96
S+ 2-*cis*-GGOH 5 μ*M*	29	30
S+ all-*trans*-FOH 10 μ*M*	37	33
S+ geraniol 5 m*M*	29	30
S+ 6,7,10,11,14,15-hexahydro-GGOH 5 μ*M*	34	46
S+ tetrahydro FOH 5 u*M*	N.A.	36

Experimental conditions: *see* **Subheading 3.2.**
N.A., not assayed.
The mean value of control (100%) without inhibitor was $104 \times 10^3 \pm 3 \times 10^3$ dpm/plate and $187 \times 10^3 \pm 10 \times 10^3$ dpm/plate for HSF and 3T3 cells, respectively.

3. To harvest the cells, place the Petri dishes on ice, remove the medium, and wash 3× with ice-cold PBS containing 1 m*M* PMSF. Scrape the cells into 1.5 mL PBS/PMSF per Petri dish, collect the resuspended cells, and centrifuge for 5 min at 200 *g*.
4. Aspirate the PBS, add 150 μL PBS/PMSF to the cell pellet, and sonicate in a bath sonicator to disrupt the cell pellet. Transfer the resuspended pellet to a 1.5-mL Eppendorf tube.
5. Add 1.3 mL of cold (–20°C) acetone, mix, sonicate, and allow to stand on ice for 15 min. Centrifuge for 5 min (13,000 *g* at 4°C) to sediment the delipidated proteins. Re-extract the protein pellet twice with 1 mL cold acetone.
6. Add 1 mL chloroform:methanol (2:1), mix, sonicate, and incubate for 30 min at 37°C. Centrifuge for 5 min (13,000 *g*, 4°C), and collect the lipid extracts.
7. Remove the residual organic solvent by evaporation, and solubilize the delipidated proteins overnight at room temperature in 100 μL 3% SDS, 62.5 m*M* Tris-HCl, pH 6.8.
8. Determine the protein content in 10 μL of sample, according to Lowry *(27)*.
9. Transfer 10 μL of the sample to a liquid scintillation vial, add 5 mL Aquamix, and measure the incorporated radioactivity.
10. Add 80 μL of sample application buffer to the remainder, and analyze an aliquot (15–80 μg of protein) by one-dimensional SDS-PAGE, according to Laemmli *(28)*, using a 12% gel.
11. After electrophoresis, wash the gel with destain (methanol–MilliQ water–acetic acid, 25:65:10) for 10 min, then stain the gel (0.1% Coomassie R250 in methanol–MilliQ–acetic acid, 40:50:10) for 20 min, and destain overnight.

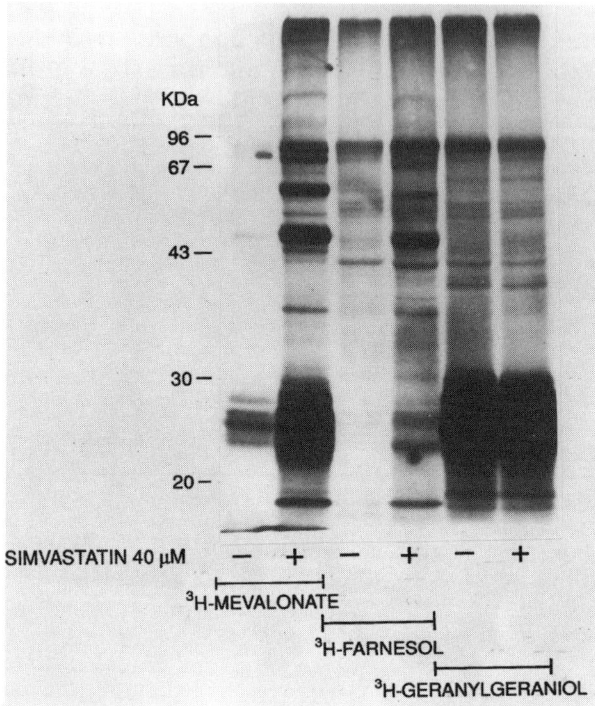

Fig. 4. The ability of simvastatin-treated HSF cells to synthesize DNA was used as a measure of their ability to traverse the cell cycle. HSF cells were seeded at a density of 3×10^5/35-mm dish in medium supplemented with 0.4% FCS, and the cultures were incubated for 72 h. Quiescent cells were incubated for 24 h in fresh medium containing 10% FCS, 40 µM simvastatin, and the indicated concentrations of unlabeled GGOH or [^{3}H]-GGOH (15 Ci/mmol). Control cells received only 10% FCS. **(Left panel)** the labeled cell lysates from each treatment group were analyzed by SDS-PAGE (*see* **Subheading 3.3., step 1**), the gel was fluorographed, and the resulting films analyzed by densitometry. The optical density reported for each treatment group represents the sum of all protein bands between 20 and 30 kDa observed in the corresponding gel lane (*see insert*). An equal amount of cell lysate (15 µg cell protein/lane) was applied to each lane. In the experiment shown, 9×10^3, 12×10^3, 42×10^3, and 142×10^3 cpm/lane were analyzed for the 0.5, 1, 2.5, and 5 µ*M* [^{3}H]-GGOH samples, for lanes left to right respectively. **(Right panel)** In a parallel experiment, DNA synthesis was determined for each treatment group. Cells were incubated as above, but with a corresponding concentration of unlabeled GGOH. After 22 h, [^{3}H]-thymidine (2 µCi/mL) was added to treated and control samples, and the incubation was continued for another 2 h. Nuclear DNA was analyzed for incorporation of [^{3}H]-thymidine (*see* **Subheading 3.2., step 7**). The incorporation of [^{3}H]-thymidine for each treatment group is reported as a percentage of the control. The mean value of control (100%) was $114 \times 10^3 \pm 6 \times 10^3$ dpm/plate.

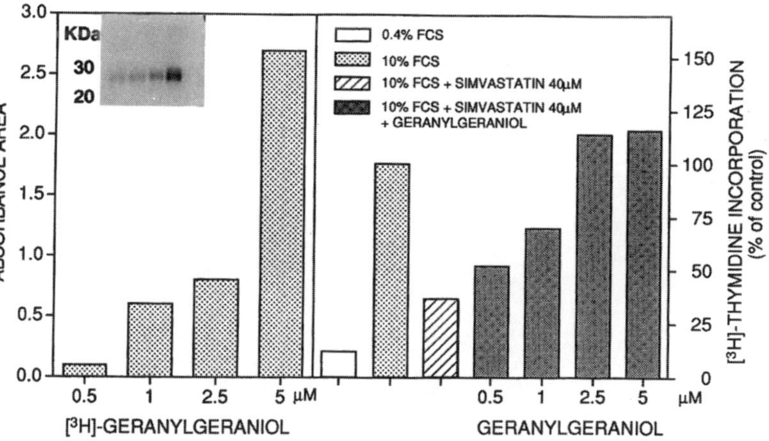

Fig. 5. Metabolic labeling of prenylated proteins in Swiss 3T3 cells after various treatments. Cells were seeded at a density of 2×10^5/dish in medium containing 1% PDS and the cultures incubated for 5 d. Quiescent cells were incubated for 20 h in fresh medium containing 10% FCS, 50 μ*M* [^{3}H]-MVA (35 Ci/mmol), 2.5 μ*M* [^{3}H]-FOH (15 Ci/mmol), or 1 μ*M* [^{3}H] GGOH (60 Ci/mmol), in the presence or absence of 40 μ*M* simvastatin. Cell pellets were delipidated, and equal amounts of cell extracts (80 μg cell protein/lane) were separated by 12% SDS-PAGE and fluorographed. In the experiment shown, 8×10^8, 2.7×10^7, 6×10^7 cpm/lane were analyzed for cells labeled with [^{3}H]-MVA, [^{3}H]-FOH or [^{3}H]-GGOH, respectively.

12. Treat the gel with Amplify for 30 min in preparation for fluorography. Wash the gel twice with water, dry, and expose to Kodak XOMAT-AR film at –70°C (**Figs. 5** and **6**).

3.3.2. High-Resolution Two-Dimensional Gel Electrophoresis of Labeled Proteins

1. In 2DE analysis, proteins are first separated on the basis of p*I*, using isoelectric focusing (IEF), followed by separation on the basis of mol mass. Therefore, differentiation of closely related proteins with similar mol masses is facile ***(29,30)***. Two different IEF methods have been developed for use in high-resolution 2DE. The one described here uses carrier ampholytes ***(31–33)*** to obtain the appropriate pH gradient in which sample proteins are focused. This system is available from ESA, Chelmsford, MA. The second IEF method uses precast gel strips containing an immobilized pH gradient made by crosslinking ampholytes into a gel matrix ***(29,34)***. This system is available from Pharmacia, Piscataway, NJ (*see* **Notes 7–9** for additional comments).
2. Prior to analysis by 2DE, cells are labeled as described in **Subheading 3.3.1.**
3. On ice, wash each 100-mm dish once with 10 mL PBS, then twice with a solution containing 10 m*M* Tris buffer, pH 7.5, 0.1 m*M* PMSF, and 1.0 μg/mL each of aprotinin, leupeptin, and pepstatin A. Drain for 45 s, and remove residual buffer.

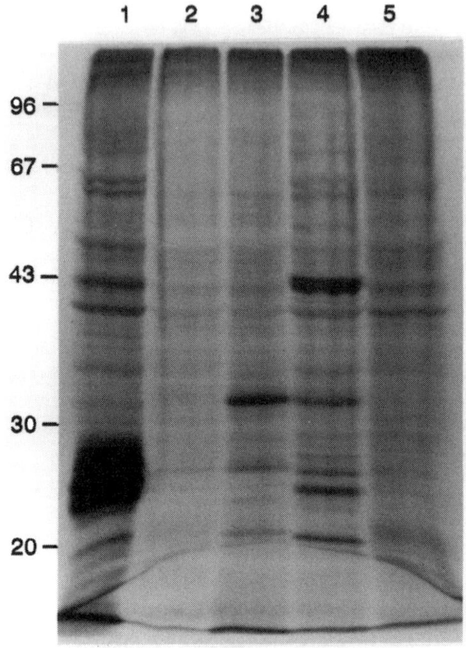

Fig. 6. Metabolic labeling of prenylated proteins in Swiss 3T3 cells after various treatments. Experimental conditions as in Fig. 5. Quiescent cells were incubated for 20 h in medium containing 10% FCS, 40 μ*M* simvastatin, and one of the following: lane 1, 5 μ*M* [^{3}H]-all *trans*-GGOH (20 Ci/mmol); lane 2, 5 μ*M* [^{3}H]-*cis, trans*-GGOH (17.5 Ci/mmol); lane 3, 5 μ*M* [^{3}H]-6,7,10,11,14,15 hexahydrogeranylgeraniol (17.5 Ci/mmol); lane 4, 5 μ*M* [^{3}H]-tetrahydrofarnesol (17.5 Ci/mmol); and lane 5, 5 μ*M* [^{3}H]-geraniol (17.5 Ci/mmol). Cell pellets were delipidated, and equal amounts of cell extract (60 μg cell protein/lane) were analyzed by SDS-PAGE on 12% gels, and gels were fluorographed. In the experiment shown, 334×10^3, 197×10^3, 247×10^3, 271×10^3, 285×10^3 cpm/lane were analyzed for lanes 1–5, respectively.

4. Add 240 μL of boiling sample buffer (28 m*M* Tris-HCl, and 22 m*M* Tris base, 0.3% SDS, 200 m*M* DTT, pH 8.0), and collect cells with a cell scraper. Transfer lysate to a 1.5-mL microcentrifuge tube, boil for 5 min, and cool on ice. This and subsequent steps are a modification of methods previously described ***(32,33)***.
5. Add 30 μL (or one-tenth vol) of protein precipitation buffer (24 m*M* Tris base, 476 m*M* Tris-HCl, 50 m*M* $MgCl_2$, 1 mg/mL DNaseI, 0.25 mg/mL RNaseA, pH 8.0) to each cell lysate sample, and incubate on ice for 8 min. Subsequent steps are performed at room temperature, unless otherwise indicated.
6. Add ~6000 cpm of the internal standard, [^{3}H]NEM-labeled soybean trypsin inhibitor (preparation described in **Subheading 3.3.3.**), to each sample (*see* **Note 10**).
7. Add 4 vol of acetone at room temperature to each lysate, and let stand for 20 min. Centrifuge the protein precipitate for 10 min at 16,000 *g*.

8. Remove the acetone phase, resuspend the precipitate in 800 µL fresh acetone, using a bath sonicator, and centrifuge as above for 5 min. Repeat this wash step once.
9. Let the pellet air dry for 3 min (avoid overdrying), before adding 30 µL of a 1:4 mixture of the SDS buffer (from **Subheading 3.2.2., step 4**) and urea solution (9.9 *M* urea; 4% Triton X-100; 2.2% ampholytes, pH 3–10; 100 m*M* DTT). Warm the sample to 37°C, and bath-sonicate to solubilize the protein (avoid overheating the sample). Centrifuge for 5 min at 16,000 *g*. Set aside 1 µL for scintillation counting and 1 µL for Bradford protein assay (*see* **Subheading 2.3.1.**).
10. Apply the sample (26 µL) onto a 1 × 180 mm-tube gel (4.1% T, 0.35% C; ampholyte pH range of 3.0–10.0). Focus for 17.5 h at 1000 V, then for 0.5 h at 2000 V.
11. Extrude the gel from the tube into equilibration buffer (300 m*M* Tris base, 75 m*M* Tris-HCl, 3% SDS, 50 m*M* DTT, and 0.01% bromophenol blue), and incubate it for 2 min before overlaying onto the second dimension SDS-PAGE gel (12.5% T, 0.27% C).
12. Perform SDS-PAGE on large format gels (220 × 240 × 1 mm) for 6 h, using the constant power mode (16 W/gel) in a prechilled tank equipped with Peltier cooling. When five gels are used in this mode, the running buffer temperature will increase 14°C over the course of 6 h when started at 0°C.
13. Fix gels for 1 h in methanol:Milli Q:acetic acid (50:40:10), then stain overnight with a solution containing 0.001% Coomassie R250 in methanol:Milli Q:acetic acid (25:65:10).
14. Immerse each stained gel in 200 mL of Amplify, and incubate for 20 min, with rocking. Without rinsing, dry the treated gel onto heavyweight blotting paper and expose to preflashed Hyperfilm MP for 7–14 d at –70°C. Exposures of >10 wk may be needed to visualize low abundance labeled proteins (**Fig. 7A,B**).

3.3.3. Preparation of the Two-Dimensional Gel Internal Standard, [^{3}H]NEM-Labeled Soybean Trypsin Inhibitor

1. Transfer 100 µL of a freshly prepared solution of 22 m*M* N-ethylmaleimide (NEM) in acetonitrile to a 0.5-mL Reacti-Vial, and add 250 µL [^{3}H]NEM (1 µCi/µL in pentane; 56 Ci/mmol). Mix, and let stand for 2 h uncapped in a fume hood to evaporate the pentane. The final volume will be ~110 µL, and the new specific activity will be 95.4 mCi/mmol.
2. Prepare 1 mg/mL solution of soybean trypsin inhibitor (STI) in 50 m*M* Tris, pH 8.5, and reduce the protein by adding DTT (2 m*M*, final concentration). Incubate overnight at 4°C.
3. Remove excess DTT by applying 300 µL of the reduced STI (in six 50-µL aliquots) to six prewashed spin columns, and centrifuge for 4 min at 1000 *g* in a countertop centrifuge. Prepare spin columns by placing 1 mL bed-volume of Bio Gel P6-DG (previously rehydrated in 50 m*M* Tris HCl, pH 7.5) into a 1-mL plastic tuberculin-syringe equipped with a plug of glass wool, followed by centrifugation for 4 min at 1000*g*. Rehydrate the spin columns by adding 1 mL of the pH 7.5 Tris buffer, and centrifuge again. Repeat this wash step 3× but do not

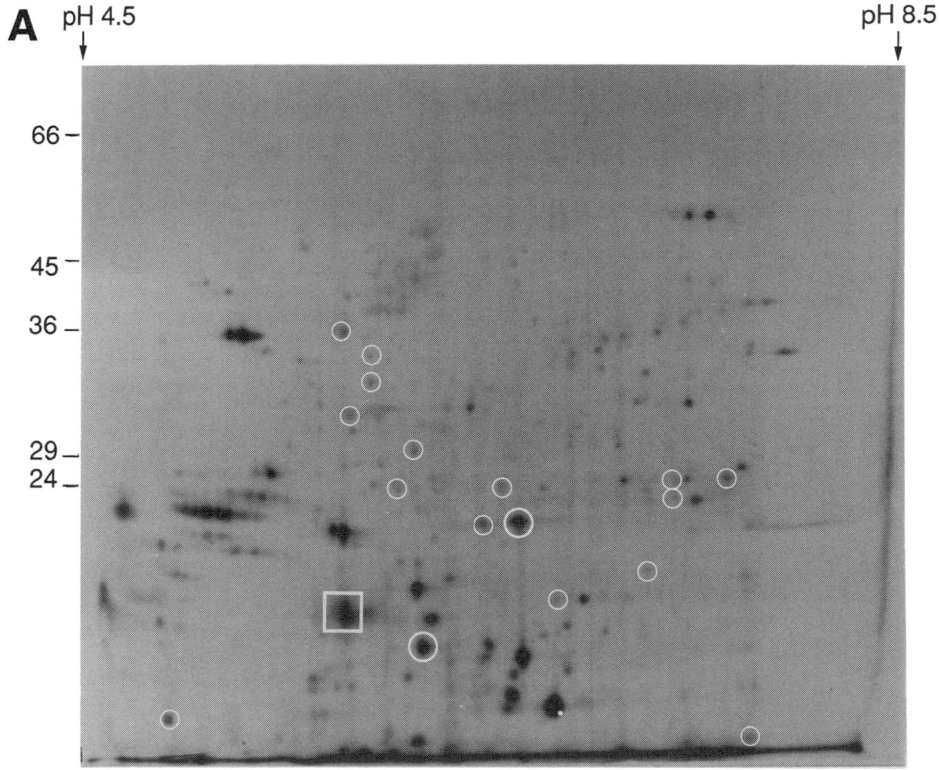

Fig. 7. Two-dimensional gel fluorographs of [^{3}H]-prenol-labeled cell-lysates. HSF cells were labeled with either 5 μm [^{3}H]-FOH or 5 μm [^{3}H]-GGOH (*see* **Subheading 3.3.1.**), lysed, delipidated and analyzed by 2DE and fluorography (*see* **Subheading 3.3.2.**). **(A)** Whole-cell lysate from [^{3}H]-FOH labeled cells that contained 385 μg protein and 170 × 10^3 cpm. Film was exposed for 23 wk. *(continued)*

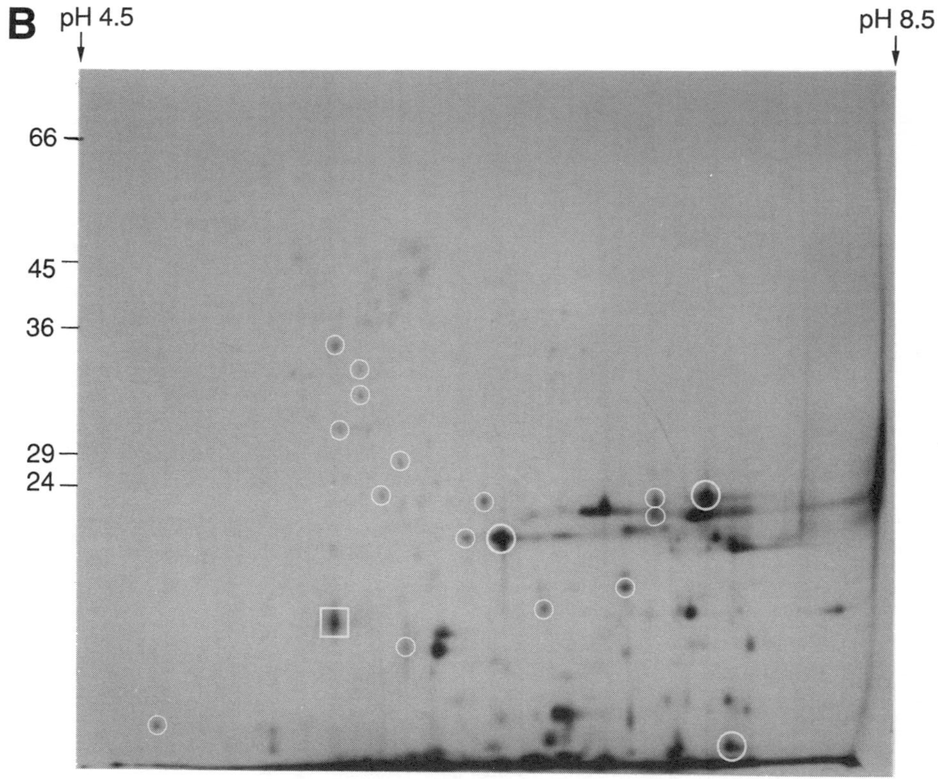

Fig. 7. *(continued)* **(B)** Whole-cell lysate from [^{3}H]-GGOH-labeled cells that contained 475 μg protein and 390×10^3 cpm. Film was exposed for 10 wk. The major proteins that incorporated both FOH and GGOH are circled. At least nine additional clusters of minor proteins display a similar ability to incorporate either prenol, but have not been circled (*see also* **Note 7**). The position of the internal standard, [^{3}H]-NEM-labeled soybean trypsin inhibitor (6250 cpm per gel; *see* **Subheading 3.3.3.**), is indicated by a square.

rehydrate the column with buffer after the last centrifugation. (In addition to removing DTT, this step also permits buffer exchange to provide the proper pH required for the NEM reaction step which follows.)

4. Pool the six column eluates (250–300 µL total volume), and add 10–12 µL of the [^{3}H]NEM solution (~24 µCi; *see* **Subheading 3.3.3., step 1**). Incubate for 6 h on ice.
5. Separate the [^{3}H]NEM-labeled STI from the unreacted [^{3}H]NEM using five or six fresh, prewashed spin columns (*see* **Subheading 3.3.3., step 3**). Pool the column eluates containing the purified [^{3}H]NEM-labeled STI (~1 µCi/nmol protein), divide into 30-µL aliquots, and store at –70°C. To make up a working stock solution, dilute the eluate 1:39 with buffer (~6000 cpm/5 µL), and use as an internal standard for 2DE gel fluorography (*see* **Subheading 3.3.2., step 6**; *see also* **Note 10**).

4. Notes

1. Since either all-*trans*-GGOH or MVA, but not all-*trans*-FOH, can completely prevent the inhibitory effect of simvastatin on DNA synthesis in cultured HSF and 3T3 cells (**Table 1**), it may be worthwhile to determine whether this effect is related to the cells' ability to incorporate labeled isoprenoid derivatives into specific proteins, or into other isoprenoid metabolites.
2. To maximize the incorporation of [^{3}H]-MVA into cellular proteins, it is necessary to block endogenous MVA synthesis with a statin such as simvastatin *(35)*. One-dimensional SDS-PAGE reveals that whole-cell homogenates incorporated [^{3}H]-MVA into at least 10 major bands, with molecular masses ranging from 21 to 72 kDa (**Fig. 5**). Intense bands of radioactivity are seen in the 21–28 kDa range, corresponding to Ras and Ras-related small GTPases *(2,3)*. In contrast, only a few protein bands are visible when simvastatin is omitted from the incubation.
3. It also is necessary to block endogenous MVA synthesis with simvastatin, for efficient incorporation of [^{3}H]-FOH into specific cellular proteins. However, this is not the case for [^{3}H]-GGOH labeling, which proceeds equally well in the presence or absence of simvastatin (**Fig. 5**).
4. In investigations of the cellular uptake of MVA and its derivatives, the authors typically found uptake of less than 0.1% of the added [^{3}H]-MVA, 2% of the added [^{3}H]-FOH, and more than 10–15% of the added [^{3}H]-GGOH. The efficient uptake of GGOH by cells permits rapid labeling of geranylgeranylated proteins. Protein bands from one-dimensional SDS-PAGE are detectable after labeling the cells for only 1–2 h, and subsequent gels require only a few days of film exposure.
5. Both [^{3}H]-GGOH labeling of proteins and the GGOH-mediated prevention of simvastatin-induced inhibition of DNA synthesis show parallel dose-response sensitivities to a similar range of GGOH concentrations (up to 2.5 µ*M*; **Fig. 4**).
6. The labeling approach described here can also be utilized for investigating natural or synthetic isoprenoid analogs (**Fig. 6**). When cells were treated with [^{3}H]-2-*cis*-GGOH (precursor of dolichols), proteins are only weakly labeled, compared to all *trans*-GGOH. Under parallel conditions, the addition of unlabeled 2-*cis*-GGOH failed to prevent the inhibitory effect of simvastatin on DNA synthesis (**Table 1**).

7. When prenylated proteins are selectively labeled with [^{3}H]-FOH or [^{3}H]-GGOH, and analyzed by one-dimensional SDS-PAGE, two subsets of proteins can be identified: a subset of 17 protein bands that incorporate [^{3}H]-FOH, and a separate subset of 12 protein bands that incorporate [^{3}H]-GGOH (*see* **Fig. 5**). However, when similarly labeled proteins are analyzed by 2DE, which allows much greater resolution, a third subset of proteins is identified. Using this approach, 82 proteins are visualized that incorporate only [^{3}H]-FOH, 34 proteins that incorporate only [^{3}H]-GGOH, and 25 proteins that can incorporate either prenol (compare **Fig. 7A,B**), although with varying efficiencies. The presence of unlabeled GGOH in the media during [^{3}H]-FOH labeling experiments or unlabeled FOH during [^{3}H]-GGOH labeling experiments, does not affect the protein composition of these subsets.
8. As mentioned above (*see* **Subheading 3.3.2., step 1**), there are primarily two different methods of IEF available for use in 2DE: one that uses soluble carrier ampholytes ***(31,32)*** and one that uses immobilized ampholytes in Immobiline™ gel-strips, discussed in **Note 10** below ***(29,30,34)***. The analyses presented here were performed using the first method, which was, until recently, the only large format system commercially available. This system is especially convenient for small analytical samples, but can also be used reproducibly for large-sample analyses ***(33)***. When 2DE is coupled with immunoanalysis, this system permits the detection of some proteins that cannot be visualized on blots from Immobiline gels (M. Aepfelbacker, personal communication; C. Farnsworth, unpublished results).
9. When large samples (>500 µg protein; e.g., cell lysates) are analyzed, the Immobiline system is currently preferable, because of advances in gel technology and sample-handling techniques. These gels are now commercially available in an extended format (0.5 × 3 × 180 mm) and come as dehydrated gel strips bonded to a rigid plastic support. This permits the gel to be rehydrated directly with dilute sample lysates. Additionally, samples containing up to 5 mg protein have also been analyzed in this way ***(36)*** without the need for concentration by techniques such as protein precipitation. The gels can accommodate a relatively large sample volume (upper limit of 350 µL vs 26 µL for the carrier ampholyte system). Another advantage of immobilized ampholyte pH gradients is that they allow the use of very narrow pH gradients (e.g., pH 5.5–6.5). This permits the separation of charge-isoforms that differ by only 0.05 pH units ***(29)***. Lastly, when 2DE is coupled with immunoanalysis and overlay assays, detailed maps of multimember protein families can be constructed ***(37)***.
10. In control experiments using lysates of unlabeled cells, recovery of the labeled internal standard, [^{3}H]-NEM STI (*see* **Subheading 3.3.3.**), following complete sample workup (*see* **Subheading 3.3.2., steps 6–9**) was 91 ± 4%, $n = 3$.

Acknowledgment

Alberto Corsini is a visiting scientist under the terms of the US (National Heart, Lung, and Blood Institute)–Italy bilateral agreement in the cardiovascular area.

References

1. Zhang, F. L. and Casey, P. J. (1996) Protein Prenylation: Molecular mechanisms and functional consequences. *Annu. Rev. Biochem.* **65,** 241–269.
2. Glomset, J. A., Gelb, M. H., and Farnsworth, C. C. (1990) Prenyl proteins in eukariotic cells: a new type of membrane anchor. *Trends Biochem. Sci.* **15,** 139–142.
3. Maltese, W. A. (1990) Posttranslational modification of proteins by isoprenoids in mammalian cells. *FASEB J.* **4,** 3319–3328.
4. Glomset, J. A. and Farnsworth, C. C. (1994) Role of protein modification reactions in programming interactions between ras-related GTPases and cell membranes. *Annu. Rev. Cell Biol.* **10,** 181–205.
5. Casey, P. J., Moomaw, J. F. Zhang, F. L., Higgins, J. B., and Thissen, J. A. (1994) Prenylation and G protein signaling, in *Recent Progress in Hormone Research*, Academic, New York.
6. Inglese, J., Koch, W. J., Touhara, K., and Lefkowitz, R. J. (1995) $G_{\beta\gamma}$ interactions with pH domains and Ras-MPK signaling pathways. *Trends Biochem. Sci.* **20,** 151–156.
7. Marshall, M. S. (1995) Ras target proteins in eukaryotic cells. *FASEB J.* **9,** 1311–1318.
8. Kato, K., Cox, A. D., Hisaka, M. M., Graham, S. M., Buss, J. E., and Der, C. J. (1992) Isoprenoid addition to Ras protein is the critical modification for its membrane association and transforming activity. *Proc. Natl. Acad. Sci. USA* **89,** 6403–6407.
9. Boguski, M. S. and McCormick, F. (1993) Proteins regulating Ras and its relatives. *Nature* **365,** 643–654.
10. Jakobisiak, M., Bruno, S., Skiersky, J. S., and Darzynkiewicz, Z. (1991) Cell cycle-specific effects of lovastatin. *Proc. Natl. Acad. Sci. USA* **88,** 3628–3632.
11. Taylor, S. J. and Shalloway, D. (1996) Cell cycle-dependent activation of Ras. *Curr. Biol.* **6,** 1621–1627.
12. Olson, M. F., Ashworth, A., and Hall, A. (1995) An essential role for Rho, Rac, and Cdc42 GTPases in cell cycle progression through G_1. *Science* **269,** 1270–1272.
13. Fenton, R. G., Kung, H., Longo, D. L., and Smith, M. R. (1992) Regulation of intracellular actin polymerization by prenylated cellular proteins. *J. Cell Biol.* **117,** 347–356.
14. Pittler, S. J., Fliester, S. J., Fisher, P. L., Keller, R. K., and Rapp, L. M. In vivo requirement of protein prenylation for maintenance of retinal cytoarchitecture and photoreceptor structure. *J. Cell Biol.* **130,** 431–439.
15. Tapon, N. and Hall, A. (1997) Rho, Rac, and Cdc42 GTPases regulate the organization of the actin cytoskeleton. *Curr. Opin. Cell Biol.* **9,** 86–92.
16. Zerial, M. and Stenmark, H. (1993) Rab GTPases in vesicular transport. *Curr. Opin. Cell Biol.* **5,** 613–620.
17. Novick, P. and Brennwald, P. (1993) Friends and family: the role of the Rab GTPases in vesicular traffic. *Cell* **75,** 597–601.
18. Raiteri, M., Arnaboldi, L., McGeady, P., Gelb, M. H., Verri, D., Tagliabue, C., Quarato, P., et al. (1997) Pharmacological control of the mevalonate pathway: effect on smooth muscle cell proliferation. *J. Pharmacol. Exp. Ther.* **281,** 1144–1153.

19. Habenicht, A. J. R., Glomset, J. A., and Ross, R. (1980) Relation of cholesterol and mevalonic acid to the cell cycle in smooth muscle and Swiss 3T3 cells stimulated to divide by platelet-derived growth factor. *J. Biol. Chem.* **255,** 5134–5140.
20. Goldstein, J. L. and Brown, M. S. (1990) Regulation of the mevalonate pathway. *Nature* **343,** 425–430.
21. Corsini, A., Mazzotti, M., Raiteri, M., Soma, M. R., Gabbiani, G., Fumagalli, R., and Paoletti, R. (1993) Relationship between mevalonate pathway and arterial myocyte proliferation: *in vitro* studies with inhibitor of HMG-CoA reductase. *Atherosclerosis* **101,** 117–125.
22. Crick, D. C., Waechter, C. J., and Andres, D. A. (1994) Utilization of geranylgeraniol for protein isoprenylation in C6 glial cells. *Biochem. Biophys. Res. Commun.* **205,** 955–961.
23. Crick, D. C., Andres, D. A., and Waechter, C. J. (1995) Farnesol is utilized for protein isoprenylation and the biosynthesis of cholesterol in mammalian cells. *Biochem. Biophys. Res. Commun.* **211,** 590–599.
24. Danesi, R., McLellan, C. A., and Myers, C. E. (1995) Specific labeling of isoprenylated proteins: application to study inhibitors of the posttranslational farnesylation and geranylgeranylation. *Biochem. Biophys. Res. Commun.* **206,** 637–643.
25. Shin-ichi, O., Watanabe, M., and Nishino,T. (1996) Identification and characterization of geranylgeraniol kinase and geranylgeranyl phosphate kinase from the *Archebacterium Sulfolobus acidocaldarius. J. Biochem.* **119,** 541–547.
26. Westfall, D., Aboushadi, N., Shackelford, J. E., and Krisans, S. (1997) Metabolism of farnesol: phosphorylation of farnesol by rat liver microsomial and peroxisomal fractions. *Biochem. Biophys. Res. Commun.* **230,** 562–568.
27. Lowry, O. H., Rosebrough, N. J., Farr, A. L., and Randall, R. J. (1951) Protein measurement with the folin phenol reagent. *J. Biol. Chem.* **193,** 265–275.
28. Laemmli, U. K. (1970) Cleavage of structural proteins during the assembly of the head of bacteriophage T4. *Nature* **227,** 680–685.
29. Görg, A., Postel, W., and Günther, S. (1988) The current state of two-dimensional electrophoresis with immobilized pH gradients. *Electrophoresis* **9,** 531–546.
30. Jungblutt, P., Thiede, B., Simny-Arndt, U. Müller, E.-C., Wittmann-Liebold, B., and Otto, A. (1997) Resolution power of two-dimensional electrophoresis and identification of proteins from gels. *Electrophoresis* **17,** 839–847.
31. O'Farrell, P. (1975) High resolution two-dimensional electrophoresis of proteins. *J. Biol. Chem.* **250,** 4007–4021.
32. Patton, W. F., Pluskal, M. G., Skea, W. M., Buecker, J. L., Lopez, M. F., Zimmermann, R., Lelanger, L. M., and Hatch, P. D. (1990) Development of a dedicated two-dimensional gel electrophoresis system that provides optimal pattern repoducibility and polypeptide resolution. *BioTech.* **8,** 518–529.
33. Lopez, M. F. and Patton, W. F. (1990) Reproducibility of polypeptide spot positions in two-dimensional gels run using carrier ampholytes in the isolelectric focusing dimension. *Electrophoresis* **18,** 338–343.

34. Görg, A., Günther, B., Obermaier, C., Posch, A., and Weiss, W. (1995) Two-dimensional polyacrylamide gel electrophoresis with immobilized pH gradients in the first dimension (IPG-Dalt): the state of the art and the controversy of vertical versus horizontal systems. *Electrophoresis* **16,** 1079–1086.
35. Schmidt, R. A., Schneider, C. J., and Glomset, J. A. (1984) Evidence for post-translational incorporation of a product of mavalonic acid into Swiss 3T3 cell proteins. *J. Biol. Chem.* **259,** 10,175–10,180.
36. Sanchez, J.-C., Rouge, V., Pisteur, M., Ravier, F., Tonella, L., Moosmayer, M., Wilkins, M. R., and Hochstrasser, D. F. (1997) Improved and simplified in-gel sample application using reswelling of dry immobilized pH gradients. *Electrophoresis* **18,** 324–327.
37. Huber, L. A., Ullrich, O., Takai, Y., Lütcke, A., Dupree, P., Olkkonen, et al. (1994) Mapping of Ras-related GTP-binding proteins by GTP overlay following two-dimensional gel electrophoresis. *Proc. Natl. Acad. Sci. USA* **91,** 7874–7878.

11

Reconstitution of Yeast Farnesyltransferase from Individually Purified Subunits

Jun Urano and Fuyuhiko Tamanoi

1. Introduction

1.1. General

Farnesyltransferase (FTase) is a heterodimeric enzyme consisting of an α- and a β-subunit involved in a crucial C-terminal modification of various proteins, including Ras ***(1–3)***. This modification by a 15-carbon isoprenyl (farnesyl) group is essential for proper localization and function of these proteins ***(4–6)***. Because the membrane localization is necessary for Ras function, inhibiting FTase will prevent this localization, which in turn will block the ability of Ras to transform cells. Therefore, the structure and function of FTase has recently been intensively studied. Many studies have characterized substrate recognition by FTase in hope of being able to design small molecules that can act as anticancer drugs by specifically inhibiting this enzyme ***(1–7)***. Recently, the structure of the rat FTase has been solved ***(8)***.

In previous studies of FTase, the enzyme is always purified as a heterodimer. Until now, the enzyme could not be purified as individual subunits and its full activity could not be reconstituted. Chen et al. ***(9)*** reported that combining crude extracts from baculovirus-infected Sf9 cells expressing individual human FTase subunits showed only a low level of activity. Another report ***(10)*** found that crude extracts from *Escherichia coli* expressing each yeast subunit can be combined to produce FTase activity, but the reconstitution was inefficient and each subunit could not be purified. Omer et al. ***(11)*** have reported a method of purifying the FTase enzyme from *E. coli* by a translationally coupled system, but showed that the expression of the individual subunits were low if expressed alone. Mayer et al. ***(12)*** using yeast FTase showed that His-tagged subunits produced inclusion bodies. Therefore, it was thought that each subunit was

From: *Methods in Molecular Biology, Vol. 116: Protein Lipidation Protocols*
Edited by: M. H. Gelb © Humana Press Inc., Totowa, NJ

relatively unstable and that both subunits needed to be cotranslated for full activity. This hindered further biochemical characterization of the individual subunits and the interaction between the two subunits. Here we present methods by which we have been able to purify the subunits of the yeast FTase. In our hands, the yeast subunits were stable and, thus, could be easily purified. We also present several procedures that allow characterization of the subunit interactions as well as individual subunit functions.

In the budding yeast, *Saccharomyces cerevisiae*, FTase is composed of Ram2, the α-subunit, and Dpr1/Ram1 (from now on referred to as Dpr1), the β-subunit ***(10,13,14)***. The α-subunit is also utilized by another prenyltransferase, geranylgeranyltransferase type I (GGTase I). Yeast GGTase I β-subunit is a protein termed Cal1/Cdc43 (from now on referred to as Cal1) ***(15)***. The yeast FTase is structurally and functionally similar to the mammalian enzyme ***(14)***. Further support for the functional similarity of the yeast and mammalian enzymes was provided by the observation that the expression of both α- and β-subunits of human FTase complemented FTase deficiency in yeast (DelVillar and Tamanoi, unpublished).

2. Materials

2.1. Purification of Individual Subunits of Yeast Farnesyl Transferase

2.1.1. Purification of GST-Dpr1

1. pGEX-DPR1 (*see* **Subheading 4.1., step 1** and **Fig. 1A**).
2. LB liquid media supplemented with 100 μg/mL ampicillin (LB + amp).
3. Isopropyl-β-D-thiogalactoside (IPTG) powder.
4. Sonication buffer (1X PBS [150 m*M* NaCl, 35 m*M* K_2HPO_4, 15 m*M* KH_2PO_4, pH 7.1 ***(16)***], 1 m*M* EDTA, 1 m*M* EGTA, 0.1% Lubrol, 0.1 m*M* DTT).
5. Elution buffer w/o glutathione (50 m*M* Tris-HCl, pH 8.0, 0.5 m*M* $MgCl_2$, 0.5 m*M* DTT).
6. Elution buffer (50 m*M* Tris-HCl, pH 8.0, 0.5 m*M* $MgCl_2$, 0.5 m*M* DTT, 8 m*M* glutathione).
7. Lysozyme.
8. Glutathione beads (Sigma, St. Louis, MO).
9. Protease inhibitors (Complete™ protease inhibitor cocktail from Boehringer Mannheim (Indianapolis, IN), 25X concentration in H_2O, kept in ice during purification).
10. Glycerol.
11. Centricon concentrator 50 (Amicon, Beverly, MA).

2.1.2. Purification of Nonfusion Dpr1

1. pCYB-DPR1 (*see* **Subheading 4.1., step 2** and **Fig. 1B**).
2. LB liquid media supplemented with 100 μg/mL ampicillin (LB + amp).

A

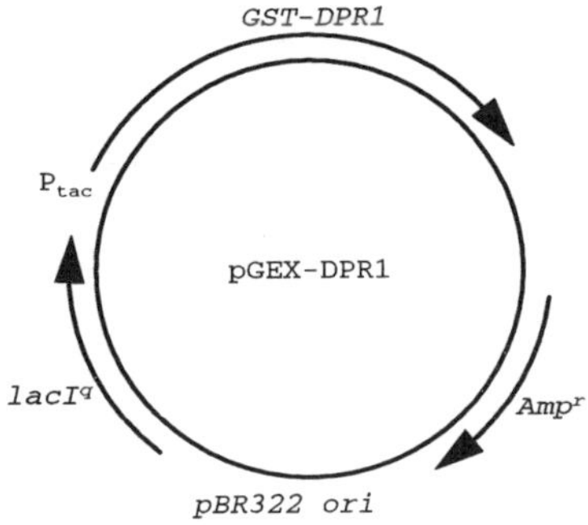

GST ... ATC GAA GGT CGT/GGG ATC GAT CCC CAG *GAA TTC* CCG GGG ATC CCG **CGA ...** ***DPR1***

B

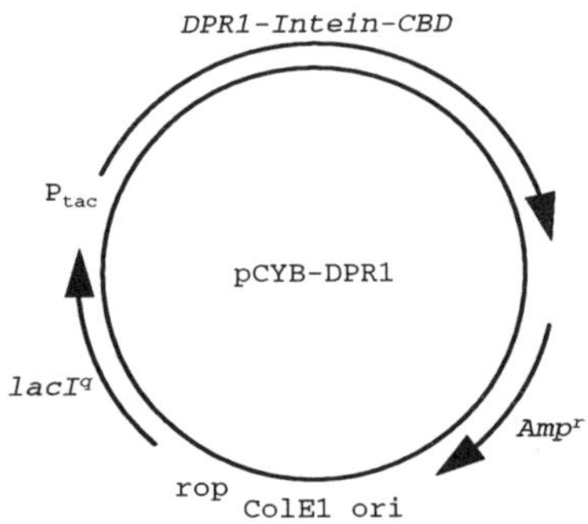

P_{tac} - CAT **ATG ...** ***DPR1*** **... TCT CCA AGT**\TGC ... Intein - CBD

C

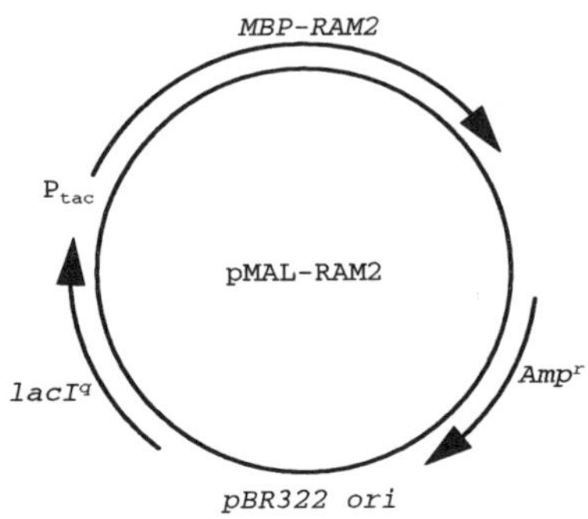

malE...ATC GAG GGA AGG/ATT TCA *GAA TTC* ATG TCG AGG AAA **ATG GAG GAG ...** ***RAM2***

Fig. 1. Expression plasmid constructs. **(A)** pGEX-DPR1 **(B)** pCYB-DPR1 **(C)** pMAL-RAM2. The nucleotide sequence of the junction is shown under each plasmid. In the case of pGEX-DPR1 and pMAL-RAM2, the codons encoding for the Factor Xa cleavage site is underlined, and the location of cleavage is indicated with a forward slash (/). In pCYB-DPR1, the location of intein-mediated cleavage is indicated with a backward slash (\). The sequence coding for *DPR1* or *RAM2* is in bold.

3. IPTG powder.
4. Cell lysis and column (CLC) buffer: 20 m*M* Tris-HCl, pH 8.0, 500 m*M* NaCl, 0.1 m*M* EDTA, 0.1% Triton X-100.
5. Cleavage buffer: 20 m*M* Tris-HCl, pH 8.0, 50 m*M* NaCl, 0.1 m*M* EDTA, 30 m*M* DTT.
6. Cleavage buffer w/o DTT: 20 m*M* Tris-HCl, pH 8.0, 50 m*M* NaCl, 0.1 m*M* EDTA.
7. Chitin beads (NEB, Beverly, MA).
8. Protease inhibitors (Complete™ protease inhibitor cocktail from Boehringer Mannheim, 25X concentration in H_2O, kept in ice during purification).
9. Glycerol.
10. Centricon concentrator 30 (Amicon).

2.1.3. Purification of MBP-Ram2

1. pMAL-RAM2 (*see* **Subheading 4.1.**, **step 3** and **Fig. 1C**).
2. LB liquid media supplemented with 100 mg/mL ampicillin (LB + amp).
3. IPTG powder.
4. Column buffer: 20 m*M* Tris-HCl, pH 8.0, 200 m*M* NaCl, 1 m*M* EDTA, 1 m*M* DTT.
5. Maltose buffer: 20 m*M* Tris-HCl, pH 8.0, 200 m*M* NaCl, 1 m*M* EDTA, 1 m*M* DTT, 10 m*M* maltose.
6. Lysozyme.
7. Amylose beads (NEB).
8. Protease inhibitors (Complete™ protease inhibitor cocktail from Boehringer Mannheim, 25X concentration in H_2O, kept in ice during purification).
9. Glycerol.
10. Centricon concentrator 50 (Amicon).

2.2. FTase Reconstitution Assay

2.2.1. Reconstitution of FTase

1. Reconstitution buffer components-stocks (20X concentration).
   0.2 *M* $MgCl_2$.
   0.1 m*M* $ZnCl_2$.
   0.1 *M* DTT.
   1 *M* Tris-HCl, pH 7.4.
2. GST-Dpr1 or Dpr1 (*see* **Subheading 3.1.1.** or **3.1.2.**).
3. MBP-Ram2 (*see* **Subheading 3.1.3.**).

2.2.2. FTase Assay

1. Reconstituted FTase (*see* **Subheading 3.2.1.**).
2. Whatman no. 3 filter paper (2-cm squares).
3. 10% TCA (~200 mL in 500 mL beakers).
4. 95% ethanol (~200 mL in 500 mL beakers).

5. Acetone (~200 mL in 500 mL beakers).
6. GST-CIIS (*see* **Subheading 4.2., step 1**).
7. ^{3}H-Farnesyl pyrophosphate (^{3}H-FPP, ~20 Ci/mmol from NEN, Boston, MA).
8. Scintillation counter.
9. Scintillation cocktail (Ecoscint™ A from National Diagnostics, Atlanta, GA).

2.3. Further Analysis of FTase Enzyme Reconstitution and Individual Subunit Functions

2.3.1. Gel Filtration Assay

1. Reconstitution buffer components—stocks (20X concentration).
   0.2 *M* $MgCl_2$.
   0.1 m*M* $ZnCl_2$.
   0.1 M DTT.
   1 *M* Tris-HCl pH 7.4.
2. GF buffer (50 m*M* Tris-HCl, pH 7.4, 10 m*M* $MgCl_2$, 5 m*M* DTT, 5 μ*M* $ZnCl_2$).
3. Sephadex G-50 medium.
4. GF column (*see* **Subheading 4.3., step 1**).
5. GST-Dpr1 or Dpr1 (*see* **Subheading 3.1.1.** or **3.1.2.**).
6. MBP-Ram2 (*see* **Subheading 3.1.3.**).
7. Scintillation counter.
8. Scintillation cocktail (Ecoscint™ A from National Diagnostics).

2.3.2. MBP-Ram2 Pull-down

1. MBP-Ram2 on amylose beads. (*see* **Subheading 4.3., step 2**).
2. Column Buffer (*see* **Subheading 2.1.3.**).
3. 2X gel-loading buffer: 20% glycerol, 10% 2-mercaptoethanol, 1% SDS, 125 m*M* Tris-HCl, pH 6.9.
4. GST-Dpr1 or Dpr1 (*see* **Subheading 3.1.1.** or **3.1.2.**).
5. Anti-Dpr1 antibody ***(14)***.

3. Methods

3.1. Purification of Individual Subunits of Yeast Farnesyl Transferase

In our purification methods, we utilized glutathione-S-transferase (GST), maltose binding protein (MBP), as well as a new intein mediated purification with an affinity chitin-binding tag (IMPACT™ I, NEB, Beverly, MA). We found that the α-subunit, Ram2, could be purified as a fusion with MBP, whereas the β-subunit, Dpr1, could be purified using GST and the IMPACT™ System.

3.1.1. Purification of GST-Dpr1

The basic protocol of the purification of GST-Dpr1 is similar to a standard GST-fusion purification protocol ***(16)***. The following is a brief protocol.

1. DH5α cells that have been transformed with pGEX-DPR1 are grown in 400 mL of LB + amp at 26°C and treated overnight (approx 16 h) with 1 m*M* IPTG. Approximately 100 mg of IPTG are added at $OD_{600} \cong 0.5$.
2. The cells are harvested and resuspended in cold sonication buffer. The cells are broken by adding 1 mg of lysozyme and protease inhibitors to 1X concentration, and then sonicating on ice with six bursts of 20-s each with 20-s intervals. The lysate is cleared by centrifuging for 30 min at 10,000*g* at 4°C.
3. The supernatant is added to cold glutathione-agarose beads (Sigma, St. Louis, MO). Another portion of protease inhibitors (to 1X concentration) is also added to ensure inhibition of proteases. (Prior to use, the glutathione beads should be swollen in cold 1X PBS and then equilibrated with sonication buffer by washing three times with 10 mL of cold sonication buffer. Protease inhibitors should be added to the last wash.)
4. GST-Dpr1 is bound to the beads by rotating at 4°C for 1–2 h.
5. Wash the beads three times with 10 mL of cold sonication buffer. Protease inhibitors should be added for the first two washes. Wash one time with elution buffer w/o glutathione.
6. Elute the protein using two 1-mL applications of elution buffer. Elution is performed at 4°C for 15 min. The eluates are pooled and then concentrated using a Centricon-50. The protein can be stored at –20°C in 50% glycerol for several months. Coomassie-stained SDS-PAGE of purified GST-Dpr1 is shown in **Fig. 2A**.

3.1.2. Purification of Nonfusion Dpr1

The following purification protocol is based on the IMPACT™ I system (NEB). The method was modified for purification using smaller amounts of beads in Eppendorf tubes instead of running columns.

1. DH5α cells that have been transformed with pCYB-DPR1 are grown and induced overnight (approx 16 h) in 400 mL of LB + amp at 26°C with 1 m*M* IPTG. Approx 100 mg of IPTG are added at $OD_{600} \cong 0.5$.
2. The cells are harvested and resuspended in 10 mL of cold CLC buffer. Protease inhibitors are added to 1X concentration, and the cells are lysed by sonicating on ice six times each with a 20-s burst followed by a 20-s interval. Clear the lysate by centrifuging for 30 min at 10,000*g* at 4°C.
3. Add the supernatant to cold chitin beads (NEB). Another portion (to 1X concentration) of protease inhibitors is also added to ensure inhibition of proteases. 500 μL of chitin beads (1 mL of a 50% slurry) should be washed and equilibrated with cold CLC buffer prior to use. Protease inhibitors should also be added to the last wash.
4. Dpr1-intein-CBD is bound to the beads by mixing at 4°C for 1–2 h.
5. The beads are washed three times with 10 mL of cold CLC buffer. Protease inhibitors are added to the first two washes. The beads are then quickly washed twice with 10 mL of cold cleavage buffer. Excess buffer is removed, and the beads are allowed to sit in the remaining buffer at 4°C overnight. This overnight

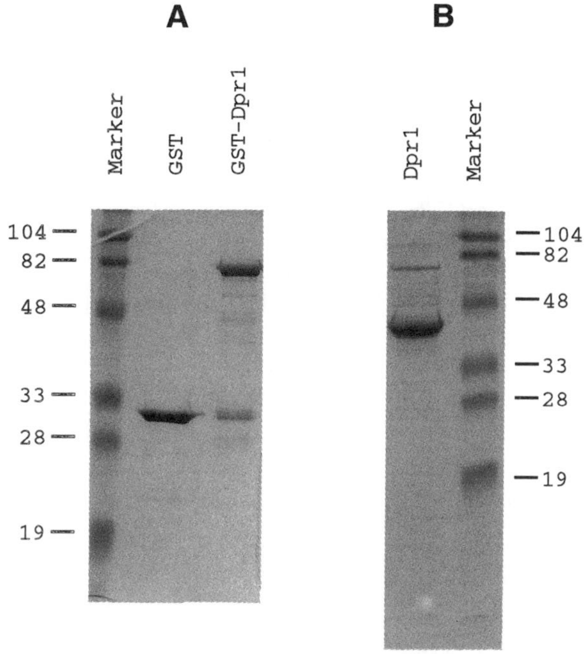

Fig. 2. Coomassie-stained SDS-PAGE of purified subunits. **(A)** Purified GST-Dpr1 runs at approx 70 kDa. This gel also shows GST alone. **(B)** Purified Dpr1 runs at the expected size of 42 kDa. **(C)** Purified MBP-Ram2 is found at roughly 80 kDa.

incubation in the presence of 30 m*M* DTT induces the intein-mediated cleavage that will release Dpr1 from the fusion protein.

6. Dpr1 is collected the following day by two 1-mL application of cold cleavage buffer w/o DTT. The two rinses are combined, and the protein is concentrated with a Centricon-30. The protein can be stored at –20°C in 50% glycerol for several months. Coomassie-stained SDS-PAGE of purified Dpr1 is shown in **Fig. 2B**.

3.1.3. Purification of MBP-Ram2

The following is a modified version of a previously described protocol *(17)*. Instead of purifying the proteins in a column, we have designed this protocol to use only a small amount of beads in an Eppendorf tube.

1. DH5α cells that have been transformed with pMAL-RAM2 are grown and induced overnight (approx 16 h) in 400 mL of LB + amp at 26°C with 1 m*M* IPTG. Approximately 100 mg of IPTG are added at $OD_{600} \cong 0.5$.
2. The cells are harvested and resuspended in 10 mL of cold column buffer. The resuspended cells are frozen.

3. The cells are then thawed, 1 mg of lysozyme is added to facilitate lysis, and protease inhibitors are added to 1X concentration. The cells are then broken by sonicating six times on ice each with a 20-s burst followed by a 20-s interval. The lysate is cleared by centrifugation for 30 min at 10,000*g* at 4°C.
4. Take the supernatant, add another portion (to 1X concentration) of protease inhibitors, and add the extract to 500 μL of cold amylose beads that has been pre-equilibrated with cold column buffer. (The amylose beads should be equilibrated with column buffer by washing three times with 10 mL of cold column buffer. In the last wash, protease inhibitors are also added.)
5. The fusion protein is allowed to bind to the beads for 1–2 h at 4°C with mixing.
6. The beads are then washed three times with 10 mL of cold column buffer. Protease inhibitors are added to the first two washes.
7. The MBP-Ram2 fusion protein is eluted into two 1-mL applications of cold maltose buffer. The elution is carried out at 4°C for 15 min. The two eluates are pooled and concentrated with a Centricon-50. The protein can be stored in 50% glycerol at –20°C for several months. Coomassie-stained SDS-PAGE of purified MBP-Ram2 is shown in **Fig. 2C**.

3.2. FTase Reconstitution Assay

3.2.1. Reconstitution of FTase

The FTase enzyme is reconstituted immediately before performing an activity assay. We have found that reconstitution using 100 nM of each subunit is sufficient to produce a significant amount of activity. However, it is advisable to alter the amount of each subunit to obtain the maximum level of reconstituted activity. According to our experience, the use of equimolar amounts of each subunit is optimal, reflecting the existence of the enzyme in vivo in a 1:1 ratio of subunits.

The final concentrations in the reconstitution reaction should be 50 m*M* Tris-HCl, pH 7.4, 10 m*M* $MgCl_2$, 5 m*M* DTT, 5 m*M* $ZnCl_2$, 1 m*M* ^{3}H-FPP, 8 m*M* GST-CIIS, and 100 nM of each subunit. All the reconstitution buffer components and the ^{3}H-FPP are mixed on ice, and brought up to volume with H_2O. The enzyme is reconstituted by simply adding both subunits. The FTase reaction is started by adding the protein substrate, GST-CIIS, and incubating at 30°C (*see* **Subheading 3.2.2.** *below*). We have found, however, that at low concentrations of subunits (10 nM), there is a slight lag in the time-course, which could be removed by a preincubation. This preincubation is carried out by incubating the reconstitution mixture at 30ºC for 10 min prior to the addition of GST-CIIS. However, at 100 nM of each subunit, no lag could be detected.

3.2.2. FTase Assay

To assess the FTase activity of the reconstituted enzyme, we usually perform a time-course, taking 10-μL samples at 0-, 5-, 10- and 20-min intervals. The FTase filter assay is performed essentially as previously described ***(13)***.

1. The reaction is started by adding the protein substrate, GST-CIIS, so that the final concentration in the reaction is approx 8 m*M*. A 10-µL aliquot is taken for the 0-min time-point and the reaction is placed at 30°C. Further 10-µL aliquots are taken at appropriate times.
2. The aliquots of the reaction mixture are spotted onto the Whatman no. 3 filter and then placed into 10% TCA. This step will precipitate the GST-CIIS and immobilize the farnesylated substrate onto the Whatman no. 3 filter.
3. After 10 min, transfer the filter to a second 10-min wash in 10% TCA. The filters are then transferred sequentially through two 10-min washes in 95% ethanol and two 10-min washes in acetone. These washes will essentially remove all free ^{3}H-FPP that may still exist in the reaction mixture. The filters should be stirred occasionally during the washes.
4. After the last wash, place the filters on a piece of aluminum foil and let dry. The filters are placed into scintillation vials, and sufficient scintillation cocktail is added to cover the filters. The amount of ^{3}H-farnesyl that was transferred to GST-CIIS is counted in a scintillation counter.

3.2.3. Example of Reconstitution Experiment

Figure 3A shows the result of a typical reconstitution experiment. The reconstituted FTase shows significant activity, and the maximum level is reached after 10 min of incubation. Dpr1 alone does not show any activity.

3.3. Further Analysis of FTase Enzyme Reconstitution and Individual Subunit Functions

3.3.1. Gel Filtration Assay

Gel filtration is utilized to assess the ability of the reconstituted FTase enzyme to complex with FPP. ^{3}H-FPP is incubated with FTase, and a Sephadex G-50 column is used to separate ^{3}H-FPP/FTase complex from free ^{3}H-FPP. The complexed ^{3}H-FPP is expected to elute at the void volume. This assay can also be modified by changing the buffer, which enables characterization of how various buffer conditions alter the ability of the reconstituted enzyme to complex with FPP. This assay can also be used with individual subunits to determine if each subunit is capable of complexing with FPP. No complex was detected when each subunit was tested alone **(Fig. 3B)**.

The following protocol is similar to that explained by Reiss *(18)*, who has utilized such an assay to identify a complex between rat FTase and ^{3}H-FPP.

1. The ^{3}H-FPP/enzyme complex is formed by incubating 25 pmol of each subunit with 100 pmol of ^{3}H-FPP in a total of 100 µL of Buffer GF for 10 min at 30°C.
2. The above mixture is applied to the Sephadex G-50 column (3 mL), which has been pre-equilibrated with buffer GF by running through roughly 10 mL of buffer GF. The applied mixture is followed by an additional 2–3 mL of buffer GF. Eighteen fractions of three drops (approx 100 µL) are collected. The column is run at 4°C.

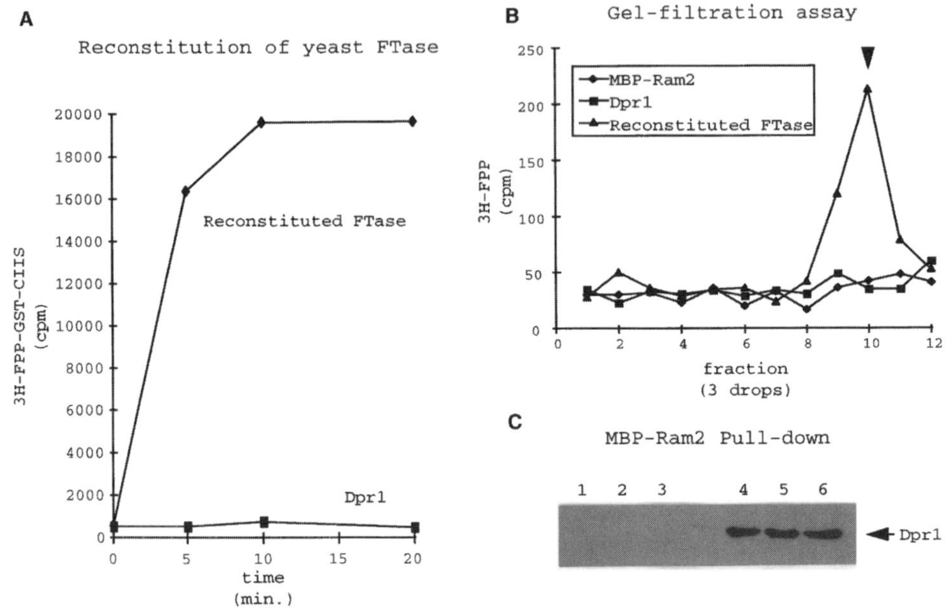

Fig. 3. Assessment of Reconstitution. **(A)** Reconstitution of FTase activity. Each subunit was used at 100 n*M*. Ten microliters of reaction mixture were spotted and counted. The reconstituted FTase (◆) shows significant activity, whereas Dpr1 alone (■) is inactive. **(B)** Gel filtration was performed using 25 pmol of each subunit for the reconstituted enzyme (▲), but 10 pmol were used for the individual subunits (Dpr1 ■, MBP-Ram2 ◆). 100 pmol of ^{3}H-FPP was used in all cases. The arrow indicates a peak representing the ^{3}H-FPP/reconstituted FTase complex. **(C)** MBP-Ram2 pull-down. Either MBP (lanes 1–3) or MBP-Ram2 (lanes 4–6) on beads were used. The FTase reaction buffer (lanes 3 and 6), or this buffer without Zn^{2+} (lanes 2 and 5), or without both Zn^{2+} and Mg^{2+} (lanes 1 and 4) was used. The arrow indicates Dpr1 visualized using anti-Dpr1 antibody.

3. Four milliliters of scintillation cocktail are added to each fraction and counted. The mixture should be well vortexed and allowed to settle (no bubbles) before counting. If the cocktail becomes cloudy, H_2O (about 100 µL) should be added until the solution clears. The complexed ^{3}H-FPP is identified as a peak at around fraction 10. The free ^{3}H-FPP will have a peak after fraction 18.

3.3.2. MBP-Ram2 Pull-down

This pull-down experiment allows us to investigate directly the interaction between the two subunits. Dpr1 or GST-Dpr1 is bound to MBP-Ram2 maintained on amylose beads, and the bound β-subunit can be visualized by a Western analysis. This method can be used to assess the buffer conditions that will allow for binding by performing the assay in various buffers containing differing amounts of salt, detergent, or divalent cations. The following protocol uses column buffer.

1. Fifty micrograms of MBP-Ram2 on beads are prepared by adding enough cold carrier amylose beads to total roughly 50 µL of beads. (The volume of the beads is increased to facilitate the washes after the binding experiment.) Twenty-five micrograms of MBP alone on beads can also be used in parallel to be used as a negative control. The beads should be resuspended in 500 µL of column buffer in an Eppendorf tube.
2. An equal molar amount of GST-Dpr1 or Dpr1 is added to the beads. The binding is done at 4°C for 1–2 h with rotation.
3. The beads are collected by centrifugation at 5000*g* for 5 min. The supernatant is discarded and the beads are washed with 500 µL of cold Column buffer. This is repeated three more times. After the last wash, the beads are resuspended using 50 µL of 2X gel-loading buffer.
4. Twenty microliters of sample is loaded per lane on an SDS-PAGE. GST-Dpr1 or Dpr1 can then be detected by Western analysis using an antibody against Dpr1. Alternatively, if GST-Dpr1 is being used, an antibody against the GST can be used.

3.3.3. Representative Results

Results of the experiments to characterize the reconstitution of FTase and individual subunit function are shown in **Fig. 3B** and **3C**. Results of gel filtration assays are shown in **Fig. 3B**. The appearance of farnesyl radioactivity at the void volume of the Sephadex G-50 column indicates that a FPP/enzyme complex is formed with the reconstituted FTase. Individual subunits, on the other hand, do not support complex formation. Another method to detect the interaction of Dpr1 and Ram2 is the MBP-Ram2 pull-down experiment. In the experiment shown in **Fig. 3C**, Dpr1 is precipitated by its interaction with MBP-Ram2 beads.

4. Notes

4.1. Purification of Individual Subunits of Yeast Farnesyl Transferase

1. The GST-Dpr1 fusion protein is expressed in *E. coli* using the plasmid pGEX-DPR1 **(Fig 1A)**. The *DPR1* gene starting from the second codon and including 3'-untranslated region was cloned into pGEX5X-3ΔB (the *Bam*HI site of pGEX5X-3 [Pharmacia] was destroyed by cutting, filling with Klenow, and self-ligating). The GST is thus fused to the N-terminus of Dpr1. The created junction introduces 15 additional amino acid residues between GST and Dpr1, including a Factor Xa cleavage site. The nucleotide sequence of this junction is shown in **Fig. 1A**.
2. pCYB-DPR1 **(Fig. 1B)** is used to express a Dpr1-Intein-CBD fusion protein in *E. coli*. The entire *DPR1* coding region (ATG to the last codon before the stop codon) was cloned into pCYB1 (NEB). The DNA fragment containing the DPR1

coding region was produced using PCR, and the sequence was confirmed by dideoxy sequencing. This leads to the Intein-CBD being fused to the C-terminal end of Dpr1. No additional amino acids are introduced. The nucleotide sequence of the junction is shown in **Fig. 1B**.

3. The plasmid used for the MBP-Ram2 purification is pMAL-RAM2 **(Fig. 1C)**. The entire coding region of *RAM2* starting from the ATG and including 3'-untranslated region was cloned into the vector pMALc2 (NEB) to produce pMAL-RAM2. The nucleotide sequence of this junction is shown in Fig. 1C. The expressed protein has the whole Ram2 sequence fused to the C-terminus of MBP. Twelve amino acid residues are introduced between MBP and Ram2, including a Factor Xa cleavage site.
4. With the use of the IMPACT™ I system, lysozyme should not be use to aid in breaking the cells. Lysozyme can cleave the chitin off the chitin beads. Also, because DTT is used to induce the cleavage, DTT is not used prior to the last washes before cleavage.
5. The Dpr1 purification methods can be modified to purify FTase enzyme as a complex. The DH5α cells are cotransformed with pBCRAM2 ***(19)*** and one of the DPR1 plasmids. The rest of the purification is done as described above.
6. For purification using GST or IMPACT™ I system, if time does not permit, once the cells are induced, the cells can be harvested, and the pellet can be frozen at –80°C. This cell pellet can then be thawed when purification is to be continued.

4.2. FTase Reconstitution Assay

1. GST-CIIS is a fusion of GST with the C-terminal 12 amino acids of the *S. cerevisiae* Ras2 protein. This fusion protein is used as the peptide substrate and is purified as previously described ***(15)***.
2. The method described here can be used to analyze the interaction between the subunits. When we used each subunit alone, we found no FTase activity, whereas the two subunits together reconstituted FTase activity **(Fig. 3A)**. We have also carried out this assay using differing ratio of the subunits to determine the subunit ratio for optimal activity as well as to determine the subunit ratio for the enzyme. In addition, the reconstitution assay is used to assess the effects that changing buffer conditions, such as salt and detergent, may have on reconstitution of active enzyme. We have found that Triton X-100 (up to 0.4%) and NaCl (up to 500 m*M*) have no significant effect on the reconstitution.

4.3. Further Analysis of FTase Enzyme Reconstitution and Individual Subunit Functions

1. For the gel filtration column, we have used a 2-mL disposable pipet. The tip is removed. One end is blocked by inserting into a Bio-Spin® disposable chromatography column (Bio-Rad) and a 3-cc syringe is inserted into the other end to function as a reservoir.
2. The MBP-Ram2 beads can be prepared by stopping the purification protocol after the last wash before elution and resuspending the beads in an equal volume of

column buffer. These beads can be maintained on ice for several weeks. The concentration of the MBP-Ram2 on beads can be calculated by running small aliquots on an SDS-PAGE gel along with BSA standards and then staining with Coomassie brilliant blue.

4.4. Conclusion/Future Prospects

This chapter is the first report of a successful purification of individual subunits of FTase and reconstitution of FTase activity. This success opens up the possibility of characterizing subunit interaction as well as each individual subunit. Preliminary results using the above assays suggest that the interaction between the subunits is resistant to high concentrations of salt (approx 60% at 500 m*M* NaCl) as well as detergent (>100% at 0.4% Triton X-100). Other salts and detergents can now be tested. In addition, the interaction does not require the addition of the divalent cations Zn^{2+} or Mg^{2+}. Furthermore, our analysis provided evidence that each subunit alone lacks FTase activity and requires the presence of the other subunit for FTase activity.

Possibilities for further characterization of the purified subunits include the use of real-time molecular interactions monitored using an optical evanescent sensor (IAsys). This could enable one to determine accurately the kinetics of the interaction. In addition, it is possible to analyze the changes in the interaction depending on the presence or lack of a substrate. Since each subunit can now be purified alone, it may be possible to determine the crystal structure of each subunit. Comparison of the structure of each subunit with the structure of the complex may enable one to detect conformational changes that may occur when the FTase complex is formed.

It will also be interesting to analyze further the function of each subunit, especially the α-subunit, Ram2. Recently, the human FTase α subunit, was identified to interact with and be phosphorylated by the cytoplasmic kinase domain of the Transforming Growth Factor-β receptor type I ***(20–23)*** Moreover, the human FTase α-subunit has also been found to be phosphorylated in response to insulin treatment of 3T3-L1 adipocytes. This phosphorylation also correlates with a slight increase in FTase activity in vitro and an increase in the pool of farnesylated p21(ras) in vivo ***(24)***. These results suggest that the FTase α-subunit may have functions different from FTase and GGTase I, or be regulated by an as yet unknown mechanism. The availability of a purified MBP-Ram2 will now allow for further characterization of Ram2 to be carried out.

Acknowledgments

We thank Wenli Yang for valuable discussions on the purification of FTase. This work was supported by National Institutes of Health Grant CA41996. J. U. is also supported by USPHS National Research Service Award GM07185.

References

1. Del Villar, K., Dorin, D., Sattler, I., Urano, J., Poullet, P., Robinson, N., et al. (1996) C-terminal motifs found in Ras-superfamily G-proteins: CAAX and C-seven motifs. *Biochem. Soc. Trans.* **24,** 709–713.
2. Sattler, I. and Tamanoi, F. (1996) Prenylation of Ras and inhibitors of prenyltransferases, in *Regulation of the RAS Signalling Network* (Maruta, H. and Burgess, A. W., eds.) RG Landes, Austin, TX, pp. 95–137.
3. Zhang, F. L. and Casey, P. J. (1996) Protein prenylation: molecular mechanisms and functional consequences. *Annu. Rev. Biochem.* **65,** 241–270.
4. Gelb, M. H. (1997) Protein prenylation, et cetra: Signal transduction in two dimensions. *Science* **275,** 1750–1751.
5. Lowy, D. R. and Willumsen, B. M. (1993) Function and regulation of RAS. *Annu. Rev. Biochem.* **62,** 851–891.
6. Khosravi-Far, R., Cox, A. D., Kato, K., and Der, C. J. (1992) Protein prenylation: key to ras function and cancer prevention? *Cell Growth Diff.* **3,** 461–469.
7. Gibbs, J. B., Oliff, A., and Kohl, N. E. (1994) Farnesyltransferase inhibitors: RAS research yields a potential cancer therapeutic. *Cell* **77,** 175–178.
8. Park, H. W., Boduluri, S. R., Moomaw, J. F., Casey, P. J., and Beese, L. S. (1997) Crystal structure of protein farnesyltransferase at 2.25 angstrom resolution. *Science* **275,** 1800–1804.
9. Chen, W. J., Moomaw, J. W., Overton, L., Kost, T. A., and Casey P. J. (1993) High level expression of mammalian protein farnesyltransferase in a baculovirus system. The purified protein contains zinc. *J. Biol. Chem.* **268,** 9675–9680.
10. He, B., Chen, P., Chen, S. Y., Vancura, K. L., Michaelis, S. and Powers, S. (1991) *RAM2*, an essential gene of yeast, and *RAM1* encode the two polypeptide components of the farnesyltransferase that prenylates a-factor and RAS proteins. *Proc. Natl. Acad. Sci. USA* **88,** 1373–1377.
11. Omer, C. A., Diehl, R. E. and Kral, A. M. (1995) Bacterial expression and purification of human protein prenyltransferases using epitope-tagged, translationally coupled systems. *Methods Enzymol.* **250,** 3–12.
12. Mayer M. P., Prestwich, G. D., Dolence, J. M., Bond, P. D., Wu, H. Y., and Poulter, C. D. (1993) Protein farnesyltransferase: production in *Escherichia coli* and immunoaffinity purification of the heterodimer from *Saccharomyces cerevisiae. Gene* **132,** 1–47.
13. Goodman, L. E., Judd, S. E., Farnsworth, C. C., Powers, S., Gelb, M. H., Glomset, J. A., et al. (1990) Mutants of *Saccharomyces cerevisiae* defective in the farnesylation of ras proteins. *Proc. Natl. Acad. Sci. USA* **90,** 2281–2285.
14. Gomez, R., Goodman, L. E., Tripathy, S. K., O'Rourke, E., Manne, V., and Tamanoi, F. (1992) Purified yeast farnesyltransferase is structurally and functionally similar to its mammalian counterpart. *Biochem. J.* **289,** 25–31.
15. Finegold, A. A., Johnson, D. I., Farnsworth, C. C., Gelb, M. H., Judd, S. R., Glomset, J. A., et al. (1991) Protein geranylgeranyltransferase of *Saccharomyces cerevisiae* is specific for Cys-Xaa-Xaa-Leu motif proteins and requires the *CDC43* gene product, but not the *DPR1* gene product. *Proc. Natl. Acad. Sci. USA* **88,** 4448–4452.

16. Ausubel, F. M., Brent, R., Kingston, R. E., Moore, D. D., Seidman, J. G., Smith, J. A., et al., (eds.) (1990) Expression and purification of Glutathione *S*-Transferase fusion proteins, in *Current Protocols in Molecular Biology*. John Wiley, New York, pp. 16.7.1–16.7.6.
17. Maina, C. V., Riggs, P. D., Grandea, A. G. III, Slatko, B. E., Moran, L. S., Tagliamonte, J. A., et al. (1988) An *Escherichia coli* vector to express and purify foreign proteins by fusion to and separation from maltose-binding protein. *Gene* **74,** 365–373.
18. Reiss, Y. (1995) Substrate interactions of proteins prenyltransferases. *Methods Enzymol.* **250,** 21–30.
19. Del Villar, K., Mitsuzawa, H., Yang, W., Sattler, I., and Tamanoi, F. (1997) Amino acid substitutions that convert the protein substrate specificity of farnesyltransferase to that of geranylgeranyltransferase type I. *J. Biol. Chem.* **272,** 680–687.
20. Miyazono, K. (1997) TGF-β receptors and signal transduction. *Int. J. Hematol.* **65,** 97–104.
21. Ventura, F., Liu, F., Doody, J., and Massague, J. (1996) Interaction of transforming growth factor-β receptor I with farnesyl-protein transferase-α in yeast and mammalian cells. *J. Biol. Chem.* **271,** 13,931–13,934.
22. Wang, T., Danielson, P. D., Li, B. Y., Shah, P. C., Kim, S. D., and Donahoe, P.K. (1996) The p21(RAS) farnesyltransferase a subunit in TGF-β and activin signaling. *Science* **271,** 1120–1122.
23. Kawabata, M., Imamura, T., Miyazono, K., Engel, M. E., and Moses, H. L. (1995) Interaction of the transforming growth factor-β type I receptor with farnesyl-protein transferase-α. *J. Biol. Chem.* **270,** 29,628–29,631.
24. Goalstone, M.L. and Draznin, B. J. (1996) Effect of insulin on farnesyltransferase activity in 3T3-L1 adipocytes. *J. Biol. Chem.* **271,** 27,585–27,589.

12

Probing the Role of H-Ras Lipidation for Signaling Functions in *Xenopus laevis* Oocytes

Thomas Dudler and Michael H. Gelb

1. Introduction

The small guanosine nucleotide binding proteins of the *ras*- gene family act as molecular switches in the transduction of many extracellular signals from cell-surface receptors to the nucleus, and are involved in a variety of cellular processes from mitosis and differentiation to apoptosis *(1,2)*. A hallmark of Ras activation is the conversion of inactive (GDP-bound) Ras to the active GTP-bound form, which then can selectively activate downstream signal transduction pathways. Ras has also been implicated in the pathogenesis of cancers, since malignant transformation in mammals is often associated with mutations in *ras* genes, which favor the GTP-bound, active form of Ras *(3)*.

A central paradigm is that Ras proteins, which are synthesized on free ribosomes, must undergo posttranslational lipidation at their C-termini before they become biologically functional. These modifications include farnesylation at a cysteine residue located four residues from the C-terminus, followed by removal of the C-terminal tripeptide and methylation of the newly exposed C-terminus *(4,5)*. Some Ras proteins (H-Ras, N-Ras, Ras2) are further lipidated by palmitoylation at one or two cysteines near the farnesylated C-terminus *(6,7)*. As a result of these lipidation reactions, Ras translocates from the cytosol to the cytoplasmic face of the plasma membrane where it is believed to engage its downstream effector pathways.

Most information regarding the sites and possible role of H-Ras palmitoylation stems from transfection studies in mammalian cell lines. These studies revealed that mutation of cysteines 181 and 184 to serine prevent palmitoylation of H-Ras in mammalian cells, suggesting that palmitoylation occurs at these cysteine residues. Additional studies revealed that palmi-

From: *Methods in Molecular Biology, Vol. 116: Protein Lipidation Protocols*
Edited by: M. H. Gelb © Humana Press Inc., Totowa, NJ

toylation is a dynamic, reversible process ***(6,8,9)***. Owing to the lack of suitable reagents (such as specific inhibitors of protein palmitoylation or depalmitoylation) to study directly the functional role of palmitoylation for H-Ras signaling, the most widely used approach is to analyze the function of palmitoylation site mutants overexpressed in transfected cell lines. Although such transfection studies can provide qualitative information, they do not readily reveal the quantitative contribution of palmitoylation to H-Ras function. Since in such transfection studies Ras is typically overexpressed at levels much higher than those present in normal cells, the contribution of a particular lipid modification to the overall biological activity of H-Ras may be underestimated. The approach described here allows the microinjection of known amounts of H-Ras into cells, thereby permitting a quantitative structure–function analysis. As will be shown, the oocyte system also permits the biochemical analysis of posttranslational processing and activation of signal transduction pathways from a small number of microinjected cells.

Most of our current knowledge regarding the role of posttranslational lipidation for the function of a lipidated protein in the context of a living cell stems from genetic approaches. This experimental strategy consists of introducing mutations into a given protein, which either adds or removes lipidation sites, and comparison of its function to the wild-type protein either by microinjection of the mutant protein into cells or by overexpression of the mutant gene in transfected cell lines. Although this approach established that lipidation is essential for Ras function, the contribution of individual lipids to the overall biology of Ras and the role of their exact chemical structure for signaling has not been forthcoming. The methodology described here has been developed to study the structure–function relationship of the H-Ras farnesyl-group for signal transduction in vivo. The approach makes use of enzymatic methods to attach structurally related farnesyl analogs onto the prenylation site of H-Ras combined with microinjection procedures to study their in vivo function in *X. laevis* oocytes. The general strategy, however, may be more widely applicable to other prenylated proteins for which a suitable cellular test system has been established.

2. Materials

2.1. Protein Preparation

1. Recombinant baculovirus expressing Glu-Glu tagged wild-type H-Ras and a mutant in which both palmitoylation sites have been mutated to serine (181/4S) (generous gift from E. Porfiri and J. F. Hancock, Onyx Pharmaceuticals, Richmond, CA).
2. 37°C water bath.
3. TBS: 10 m*M* Tris, pH 7.5, 150 m*M* NaCl.

4. 10% Triton X-114 in TBS:
   a. Weigh out 3 g of Triton X-114 (Sigma) in a 50-mL Falcon tube, and top off with ice-cold TBS; allow to dissolve overnight at 4°C on a rocking platform.
   b. Place tube in a 37°C water bath until the solution becomes cloudy (approx 5 min), and spin in a clinical centrifuge at approx 300*g* for 5 min at room temperature (allow to stop without brakes).
   c. Discard upper (aqueous) phase, and restore volume to 50 mL with ice-cold TBS.
   d. Mix to obtain a clear solution (if solution is still cloudy, cool mixture on ice until it becomes clear), then repeat **steps b** and **c** three more times.
   e. Discard the final upper phase; the final detergent phase (approx 20 mL) contains 10–11% Triton X-114 (this Triton solution can be stored for up to 2 wk at 4°C).
5. Anti-Glu-Glu MAb (kindly supplied by E. Porfiri) immobilized on protein-G Sepharose (Sigma) according to the published procedure ***(10)***.
6. GTP-γS (Sigma, 1 m*M* in water, store at –80°C).
7. 4X Nucleotide exchange buffer: 15 m*M* Tris, pH 7.4, 40 m*M* EDTA, 10 m*M* $MgCl_2$.
8. 1 m*M* DTT (store at –20°C in aliquots).
9. 1 *M* $MgCl_2$.
10. Ras buffer: 20 m*M* Tris-HCl, pH 7.4, 100 m*M* NaCl, 4 m*M* $MgCl_2$, 0.1 m*M* DTT (store at -20°C in aliquots).
11. Ultrafiltration device (Centricon-10, Amicon).
12. Refrigerated centrifuge with SS-34 rotor.

2.2. Oocyte Culture and Microinjection

1. Large, mature *X. laevis* females, oocyte-tested, from Nasco or *Xenopus I.*
2. 0.25 % MS-222 (Sigma) in tap water.
3. Surgical scissors, blunt forceps and suture.
4. OR-2 medium: 82.5 m*M* NaCl, 2 m*M* KCl, 1 m*M* $MgCl_2$, 5 m*M* HEPES, pH 7.6 (prepare a 10X stock solution, filter-sterilize, and store at 4°C; make fresh 1X medium for every experiment).
5. Modified Barth saline (MBS): 88 m*M* NaCl, 1 m*M* KCl, 2.4 m*M* $NaHCO_3$, 15 m*M* HEPES, pH 7.6, 0.3 m*M* $Ca(NO_3)_2$, 0.41 m*M* $CaCl_2$, 0.82 m*M* $MgSO_4$, 50 μg/mL gentamycin (prepare a 5X stock solution without antibiotic, filter-sterilize, and store at 4°C; antibiotic is added to 1X medium, which can be kept at 4°C for up to a week).
6. Thirty milliliters of collagenase (Sigma; cat. no. C 0130) at 1.5 U/mL in OR-2 (dissolve immediately before use).
7. Shortened Pasteur pipet: trim a Pasteur pipet so that the inner diameter is approx 2–3 mm, and polish the trimmed end in a gas flame.
8. Stereo microscope (Wild).
9. 18–20°C incubator.
10. Drummond microdispenser mounted on Narishige Micromanipulator.

11. Drummond glass capillaries (Drummond 3-000-203-G/X).
12. Vertical micropipet puller.

2.3. Oocyte Extract Preparation and Kinase Assays

Oocyte lysis buffer:

60 m*M* β-glycerophosphate, pH 7.4, 20 m*M* EGTA, 15 m*M* $MgCl_2$, 1 m*M* Na_3VO_4, 1 m*M* NaF, 1 m*M* benzamidine, 5 μg/mL leupeptin, and 1 m*M* PMSF (store at –20°C in aliquots without protease inhibitors; add inhibitors immediately before use)

MAPK assay:

1. MBP: myelin basic protein (Sigma, cat. no. P 1891) 2 mg/mL in water (store in aliquots at –80°C).
2. M1 mix: 450 m*M* β-glycerophosphate, pH 7.4, 175 m*M* $MgCl_2$, 25 m*M* EGTA, 9 m*M* DTT, 3 m*M* Na_3VO_4 (store in aliquots at –20°C, and discard after use).
3. Nonlabeled ATP, 10 m*M* in water (store in aliquots at –80°C, and discard after use).
4. [γ-^{32}P] ATP (NEN, 1000–4000 Ci/mmol).
5. PKI: 0.5 m*M* heat-stable inhibitor of cAMP-dependent protein kinase (Sigma, cat. no. P 1891) in water (store at –80°C).
6. CMZ: 1.25 m*M* calmidazolium (Sigma) in water (store at –80°C).
7. Dry heating block accommodating 0.5-mL microfuge tubes.
8. p-81 Phosphocellulose paper (Gibco-BRL): cut 4 × 6 disk sheets into strips of 4 disks.
9. 600-mL beaker fitted with mesh-wired basket and magnetic stir bar below basket.
10. 150 m*M* H_3PO_4.

Maturation-promoting factor (MPF) assay:

1. Histones (type III, calf thymus, Sigma cat. no. H 4524) 5 mg/mL in water (store in aliquots at –80°C).
2. 2X Hot mix: 40 m*M* HEPES pH 7.5, 10 m*M* mercaptoethanol, 20 m*M* $MgCl_2$, 200 μ*M* ATP (S. A. ~ 1 Ci/mmol), 10 μ*M* PKI (store buffer aliquots at –80°C; add hot ATP and PKI from stock solution immediately before use).
3. 4X SDS-PAGE loading buffer: 200 m*M* Tris, pH 6.8, 8% SDS, 40% glycerol, 0.025% Bromophenol blue, 200 m*M* DTT (added freshly).
4. 12.5% SDS-polyacrylamide gels.
5. Bio-Rad model 583 Gel dryer.
6. Kodak X-O-mat X-ray films.

2.4. Posttranslational Processing and Membrane Translocation

Radiolabeling studies:

1. BSA-treated microcentrifuge tubes: Pretreat microcentrifuge tubes with 1 mg/mL BSA in water for 1 h at room temperature; remove BSA solution, and wash carefully with water, and then with MBS.

2. [^{3}H] FPP: Farnesyl-pyrophosphate (ARC; 15 Ci/mmol; 1 mCi/mL; 66 μ*M*).
3. [Methyl-^{3}H] methionine (NEN, 70 Ci/mmol).
4. [^{3}H] palmitic acid (NEN, 60 Ci/mmol).
5. 5 m*M* NH_4HCO_3.
6. Speed-Vac.
7. Bath sonicator.

Triton X-114 phase partitioning:

1. 37°C Water bath.
2. TBS: 10 m*M* Tris, pH 7.5, 150 m*M* NaCl.
3. 10% Triton X-114 in TBS (*see* **Subheading 2.1.**, **step 4**).

Oocyte fractionation:

1. T + 10 buffer: 20 m*M* Tris, pH 7.6, 100 m*M* NaCl, 50 m*M* KCl, 10 m*M* Mg acetate, 10% sucrose (w/v) (store at –20°C).
2. T + 20 buffer: 20 m*M* Tris, pH 7.6, 100 m*M* NaCl, 50 m*M* KCl, 10 m*M* Mg acetate, 20% sucrose (w/v) (store at –20°C).
3. 10% Triton X-100 in water.
4. Microcentrifuge tubes with pestle (Kimble).
5. Polyallomer ultracentrifuge tubes (Beckman).
6. Tabletop ultracentrifuge (Beckman) with TI-45 rotor.

2.5. Immunoprecipitation

1. IP buffer: 10 m*M* Tris, pH 7.4, 150 m*M* NaCl, 1 m*M* EDTA, 1 m*M* EGTA, 0.2 m*M* Na_3VO_4, 1% Triton X-100, 0.5% NP-40 (store at –20°C).
2. Anti-Ras Sepharose: 50% slurry of Ab-1 Sepharose (Oncogene Sciences).
3. Rocking platform.

2.6. Immunoblotting

1. 10 or 15% SDS-polyacrylamide gels.
2. Minitransblot assembly (Bio-Rad).
3. Nitrocellulose membranes (Hybond; Amersham).
4. Sheep antimouse IgG-horseradish peroxidase conjugate (Amersham).
5. ECL detection kit (Amersham).
6. Hyperfilm (ECL).

Ras and MAPK immunoblots

1. Wash buffer: 10 m*M* Tris, pH 7.5, 100 m*M* NaCl containing 0.1%Tween-20.
2. Blocking buffer: 5% nonfat milk in wash buffer.
3. Anti-Ras murine MAb (Transduction laboratories, cat. no. R02120).
4. Anti-ERK2 murine MAb (Transduction laboratories, cat. no. E16220).

Phosphotyrosine Immunoblot:

1. Wash buffer: 10 m*M* Tris, pH 7.5, 100 m*M* NaCl containing 0.5%Tween-20.
2. Blocking buffer: 2% BSA (Sigma, cat. no. A 7906) in wash buffer.

3. Antiphosphotyrosine antibody (PY-20; Transduction laboratories).

2.7. In Vitro Incorporation of Farnesyl Analogs into H-Ras

1. PFT: Recombinant protein farnesyltransferase purified from Sf9 as described *(10,11)*.
2. H-Ras^{V12}: Activated human H-Ras carrying the Val-12 mutation is expressed in *E. coli* using the pMG27/T24 plasmid (kind gift from Prof. F. Tamanoi, UCLA) and purified as previously described *(11)* (*see* **Note 5**). Flash-freeze the final H-Ras preparation (approx 80 μM) in liquid nitrogen, and store at –80°C.
3. Farnesyl pyrophosphate (FPP) analogues: A detailed description of the synthesis of the pyrophosphates of the farnesyl analogs geranyl (G), tetrahydrofarnesyl (THF) and (E)-3-methyl-2-dodecenyl (MD) is given elsewhere *(11)*. All FPP analogues are suspended at 1 m*M* in 5 m*M* NH_4HCO_3.
4. 2X PFT buffer: 60 m*M* Tris, pH 7.4, 100 m*M* NaCl, 10 m*M* $MgCl_2$, 20 μM $ZnCl_2$, 1 m*M* DTT (freshly added).
5. GTP: 1 m*M* solution in water.
6. 10X Nucleotide-exchange buffer: 100 m*M* Tris, pH 7.4, 100 m*M* EDTA, 25 m*M* $MgCl_2$, 10 m*M* DTT.
7. 1 M $MgCl_2$.
8. 2% Octyl-glucoside in water.
9. Ras buffer: 20 m*M* Tris pH: 7.4, 100 m*M* NaCl, 4 m*M* $MgCl_2$, 0.1 m*M* DTT, 0.1% octyl-glycoside.
10. Ten Biogel P-6 (Bio-Rad) spin columns: prepare about 12 mL of a 50% slurry of P-6 Biogel in Ras buffer, and transfer 1 mL of slurry into a 1-mL tuberculin syringe fitted with a glass wool plug. Spin for 5 min at approx. 1000*g* in a swing-out rotor in a clinical centrifuge at room temperature to remove excess buffer (bed volume will be about 0.5 mL). Equilibrate resin by adding 100 μL of Ras buffer and spinning for 2 min as above. Repeat this step twice. No buffer should be left on top of the resin before loading the sample.
11. Five "passivated" Microcon-10 (Amicon): Fill sample chamber with a solution of 5% Tween-20 in water, and allow to stand for about 1 h at room temperature. Rinse thoroughly with deionized water, and then with Ras buffer. Fill the device with Ras buffer and spin for about 5 min at 4°C in the concentration mode to probe for potentially leaky devices. Keep device filled with Ras buffer, and pour out excess Ras buffer just before applying sample for concentration.

3. Methods

3.1. Preparation of Activated Wild-Type H-Ras and Palmitoylation Site Mutant

For the studies described herein, an epitope-tagged form of wild-type H-Ras and the corresponding mutant in which the palmitoylated cysteine residues have been mutated to serine (H-Ras 181/4S) were expressed in Sf-9 cells, and the nonlipidated form was isolated from the aqueous phase of Sf-9 cell lysates following Triton X-114 phase partitioning. A detailed isolation and purification procedure for these proteins is given elsewhere (***12***; *see* also **Note 1**). The H-Ras concentration of the purified preparation (typically 5–20 μ*M*) was determined by measuring the incorporation of [^{3}H]-FPP catalyzed by protein farnesyltransferase (PFT) as described ***(10)***, and protein samples were stored in aliquots at –80°C.

The active form of wild-type and mutant H-Ras was prepared as follows. All steps are carried out at 4°C unless noted otherwise:

1. In a microfuge tube on ice, mix 2–4 nmol of purified H-Ras (2–4 μ*M* final) with 250 μL GTPγS (250 μ*M* final) and 250 μL of 4X nucleotide exchange buffer, and adjust the volume to 1 mL with water.
2. Incubate for 20 min at 30°C.
3. Stop exchange reaction by placing tubes on ice and add 30 μL 1 *M* $MgCl_2$ (30 m*M* final).

To remove excess nucleotide and concentrate the sample proceed as follows:

4. Transfer mixture into ultrafiltration device (Centricon-10), and concentrate sample to about 50 μL by centrifugation for 45 min at 5000*g* in a fixed-angle rotor.
5. Wash by adding 1 mL of ice-cold Ras buffer and performing the concentration step above.
6. Repeat this washing step three more times and concentrate the H-Ras preparations to a final volume of about 50 μL.
7. Spin out concentrated H-Ras sample into collection tube.
8. Add another 20 μL of Ras buffer into the sample chamber of the centricon, vortex, spin out sample, and combine with first eluate (*see* **Note 6**).
9. Freeze in 10-μL sample aliquots in 0.5-mL Eppendorf tubes, and store at –80°C.
10. Determine H-Ras concentration in these sample aliquots by measuring the incorporation of [^{3}H]-FPP catalyzed by PFT as described ***(10)*** (typically 30–50 μ*M*).

3.2. Oocyte Culture and Microinjection

All oocyte isolation steps are carried out at room temperature, and oocytes are subsequently cultured in MBS at 18–20°C. Procedures to anesthetize animals and to remove ovarian lobes surgically from *X. laevis* females have been described in detail ***(13)***. This reference also provides the technical details of the microinjection procedure. Follow these instructions except for the steps listed below.

1. Cut isolated ovarian lobes in MBS medium into small (pea-size) clumps with scissors.
2. Transfer dissected ovarian lobes to a 15-mL falcon tube.
3. wash tissue with 10 mL of OR-2 medium by gentle end-over-end tumbling (three to five times); allow tissue to settle (about 5–10 s), then aspirate excess medium.
4. Repeat washing step 5–10 times.

5. Aspirate excess OR-2 medium, and fill falcon tube to about 14 mL with collagenase solution (1.5 U/mL in OR-2).
6. Place Falcon tube on a rocking platform, and gently rock for 1 h at room temperature.
7. Aspirate collagenase solution, and wash cells five times with OR-2.
8. Digest tissue for another hour with collagenase by repeating **steps 5** and **6**.
9. Aspirate collagenase, wash cells with OR-2 until the buffer is clear (about 10 times), then twice with MBS, and pour cells into a 60-mm petri dish.
10. Under the microscope, identify large (>1 mm diameter) stage V–VI oocytes with homogeneous pigmentation, and transfer them to another MBS-containing Petri dish using a shortened Pasteur pipet.
11. Allow oocytes to recover overnight, and discard cells with pigment diffusion or irregularities before microinjection.
12. Scratch a grid into the bottom of a Petri dish with an injection needle, fill it with MBS, and place it under a stereo-microscope.
13. With a shortened Pasteur pipet align 20–40 oocytes/experimental group in this Petri dish.
14. Microinject oocytes with 50 nL/cell of appropriately diluted protein solution (*see* **Note 2**), transfer cells with a shortened Pasteur pipet to a new Petri dish containing MBS, and continue incubation at 18–20°C.

3.3. Oocyte Maturation, Extract Preparation and Determination of Kinase Activation

Maturation and extract preparation

1. As a positive control for maturation, incubate 10–20 oocytes in MBS medium containing 10 μg/mL of progesterone.
2. At regular time intervals (1–2 h, starting 3 h after microinjection), monitor oocytes for geminal vesicle breakdown (maturation) by the appearance of a white spot on the animal (pigmented) pole of the oocyte. Details of this process as well as additional methods to confirm maturation are presented elsewhere ***(14,15)***.
3. At selected time-points after microinjection, randomly collect 5 oocytes/group using a shortened Pasteur pipet and transfer to an eppendorf tube. Remove excess medium, flash-freeze in liquid nitrogen, and store at –80°C until used.
4. Thaw oocytes on ice and homogenize in 10 μL/cell of lysis buffer by pipeting the cells 15–20 times through a yellow pipetter tip; vortex briefly, and allow to sit on ice for 15 min. Freeze a portion (10–20 μL) of this crude lysate for membrane fractionation and Triton phase partitionning studies (*see* **Subheading 3.5.** and **Subheading 3.4.**, last section); proceed with the remainder as follows.
5. Pellet insoluble matter at 12,000*g* for 15 min at 4°C.
6. Promptly remove the cleared extract, and set in a fresh tube on ice. Cleared extract not used immediately can be flash-frozen in liquid nitrogen and stored at –80°C for later experiments.

MAPK assay: Since this is a timed assay it is best to stagger the initiation of reactions by 15 s. Keep all reagents and assay mixtures on ice until reactions are initiated.

1. For 24 kinase assays, prepare 260 µL 3X Hot assay mix: In a 0.5-mL Eppendorf tube, combine 45 µL M1, 7.8 µL nonlabeled ATP, 3 µL PKI, 6 µL CMZ, and 26 µCi [γ-^{32}P] ATP, and adjust the volume to 260 µL with water.
2. Set up reactions in 0.5-mL Eppendorf tubes by mixing 12.5 µL of water with 5 µL of MBP substrate solution followed by the addition of 2.5 µL cleared oocyte extract; for each assay, include a control reaction without substrate (17.5 µL water + 2.5 µL extract).
3. To initiate the reaction, add 10 µL of 3X Hot mix, vortex, and set tube in a 30°C heating block; initiate next reaction staggered by 15 s.
4. After an incubation period of 20 min, terminate first reaction by spotting 20 µL of the reaction mixture onto a phosphocellulose paper disk. Process subsequent reaction mixtures staggered by 15-s time intervals.
5. After spotting the first four reaction mixtures onto the 4 disks of the first paper strip, transfer paper strip into mesh-wired wash breaker containing 450 mL 150 m*M* H_3PO_4. Proceed with the next four reaction mixtures as above.
6. After terminating all reactions, wash off excess ATP by setting the beaker on a stir plate and stir for 15 min.
7. Decant wash solution into shielded decay container, add another 450 mL of 150 m*M* H_3PO_4, and stir for another 15 min.
8. Repeat this washing step three more times, then rinse phosphocellulose paper briefly with ethanol, and allow to air-dry.
9. Remove individual disks from paper strip with forceps, transfer to a scintillation vial, and count radioactivity after adding 0.5 mL of scintillation fluid (BioSafe II, Beckmann).

For each extract, the net MAPK activity is obtained by subtracting the background (cpm in reaction without MBP) from the cpm of the complete kinase reaction. The stimulation index (fold MAPK activation) is obtained by dividing the net cpm of Ras-injected cells by the net cpm of buffer injected cells.

MPF assay:

1. For each assay, mix 7.5 µL of water with 15 µL of 2X Hot mix and 5 µL Histone in a 0.5-mL Eppendorf tube on ice.
2. Add 2.5 µL oocyte extract, vortex briefly, and set tube in a 30°C heating block.
3. After 15 min of incubation, stop reaction on ice, add 8 µL of 4X SDS-PAGE loading buffer, and vortex briefly.
4. Boil samples for 5 min, spin briefly, and load sample onto 10% SDS-PAGE. Include one lane with prestained mol-wt standards.
5. Run gels at 200 V for 50 min, and fix with 40% methanol, 10% acetic acid for 1 h.
6. Dry gels for 1 h at 80°C and expose to film for 1–4 h at room temperature. MPF activity will phosphorylate Histone H1, resulting in a strong radiolabeling of a ~35 kDa protein.

3.4. Posttranslational Processing

Analysis of farnesylation by coinjection of H-Ras and [^{3}H] FPP:

1. Concentrate [^{3}H] FPP to 0.2 m*M* for comicroinjection as follows: set 30 µL of [^{3}H] FPP supplied as a 66-µ*M* solution in ethanol/0.25 M NH_4HCO_3 (1:1) in an Eppendorf tube, and place in a Speed-Vac for 30 min to remove ethanol. Take up the residue in 0.25 mL of 5 m*M* NH_4HCO_3, freeze, and lyophilize. Repeat this step two more times. Dissolve in 10 µL 5 m*M* NH_4HCO_3 to obtain a final concentration of 0.2 m*M*.
2. Combine 1 µL of 0.2 m*M* [^{3}H] FPP with 1 µL of a 20-µ*M* solution of mutant or wild-type H-Ras on a piece of parafilm. Mix by pipeting 5–10 times.
3. Take up mixture in injection needle, inject 30 oocytes with 50 nL of this mixture and continue oocyte culture in MBS medium.
4. After 2.5 to 24 h of culture, randomly harvest five oocytes and prepare cleared extracts as described above (*see* **Subheading 3.3.**) but using lysis buffer supplemented with 1% Triton X-100.
5. Immunoprecipitate H-Ras from these extracts as described below.

Methylation

1. Prepare 250 µL labeling medium containing 0.4 mCi/mL of [Methyl-^{3}H] methionine as follows: Transfer 100 µCi [Methyl-^{3}H] methionine into a BSA-treated Eppendorf tube, and remove solvent in a Speed-Vac; resuspend in 250 µL MBS by vortexing extensively (~1 min), and verify that most of the methionine has been solubilized by submitting 0.5 µL of the sample to scintillation counting.
2. Microinject 20 oocytes/group with 1 pmol of H-Ras.
3. Immediately after injection, transfer oocytes to a BSA-treated eppendorf tube. Remove excess MBS medium and add 6 µL/oocyte of labeling medium. Tap tube gently to mix medium, and lay tube flat on the bench to avoid packing of the oocytes.
4. After selected incubation periods, collect 5 oocytes/group, remove excess labeling medium and freeze in liquid nitrogen.
5. Prepare oocyte lysates using lysis buffer containing 1% Triton X-100, and immunoprecipitate H-Ras as described below.

Palmitoylation:

1. Prepare 0.4 mL of labeling medium containing 2 mCi/mL of [^{3}H]palmitate as follows: To a BSA-treated glass vial add 0.8 mCi of [^{3}H]palmitic acid, and remove solvent in a Speed-Vac. Add 0.4 mL of MBS, and solubilize [^{3}H]palmitate by vortexing and sonication (~1 min each). Verify that most of the palmitate has been solubilized by submitting 0.5 µL to scintillation counting.
2. Microinject 20 oocytes/group with 1 pmol of H-Ras, and transfer to BSA-treated microfuge tube containing nonlabeled MBS medium.

3. After 4 h of incubation in nonlabeled MBS medium at 18–20°C, remove excess MBS, and add 10 µL/oocyte of [^{3}H]palmitate-containing medium.
4. Tap tube gently to mix medium, and lay tube flat on the bench to avoid packing of the oocytes.
5. After a labeling period of 10–120 min, harvest 5 oocytes/group, remove excess labeling medium, and freeze in liquid nitrogen.
6. Prepare oocyte lysates using lysis buffer containing 1% Triton X-100 and immunoprecipitate H-Ras as described below.

Triton X-114 phase partitioning:

1. Mix 10 µL crude oocyte lysate (**step 4, Subheading 3.3.**) with 100 µL of 1% Triton X-114 in TBS, and allow to sit on ice for 10 min with occasional mixing also (*see* also **Note 3**).
2. Remove insoluble debris by centrifugation (10 min at 10,000*g*) and set cleared lysates into a fresh tube.
3. To separate the mixture into aqueous (a) and detergent (d) phases, set tube in a 37°C water bath for 2 min, and then centrifuge for 2 min at 10,000*g* at room temperature.
4. Carefully remove the upper (a) phase, and set into a fresh tube.
5. Chill tubes containing separated a and d phase on ice, and adjust both phases to 1% Triton X-114 and 100 µL final volume (by adding 90 µL TBS to d phase and 10 µL 10% Triton to a phase).
6. Mix 10 µL of a and d phase with 10 µL 2X LaemmLi buffer, boil for 2 min, and subject to SDS-PAGE/immunoblotting analysis as described below.

3.5. Oocyte Fractionation

All operations are carried out on ice or at 4°C.

1. Mix 10 µL of crude oocyte extract (from **step 4, Subheading 3.3.**) with 50 µL of T + 10 buffer and homogenize in an eppendorf tube using a pestle.
2. Sonicate briefly in a bath sonicator and layer the homogenate onto 100 µL T + 20 buffer in a polyallomer tube.
3. Spin for 1 h at 20,000*g* using a TI-45 rotor in a tabletop ultracentrifuge.
4. Carefully remove 50 µL of the top layer (s-phase), and set in a fresh tube; remove the interphase and bottom layer covering the membrane pellet, and discard.
5. Resuspend membrane pellet in 54 µL T + 10 buffer with a pestle, then add 6 µL 10% Triton X-100, vortex, and allow to sit on ice for 20 min.
6. Spin down insoluble matter for 15 min at 10,000*g*, remove 50 µL Triton extract (p-phase), and set in a fresh tube.
7. Mix 6 µL of s-phase and p-phase of each oocyte extract with 20 µL 1X SDS-PAGE buffer, boil for 2 min, and load samples onto 15% SDS-PAGE.
8. Run gels for 60 min at 200 V, blot proteins onto nitrocellulose membranes, and reveal the presence of Ras in the s-phase and p-phase by anti-Ras Western blotting as described below.

3.6. Immunoprecipitation

1. For each immunoprecipitation assay, mix 450 µL IP buffer with 30 µL cleared oocyte extract (3 oocyte equivalent) and 15 µL anti-Ras Sepharose in a 1.5-mL microfuge tube on ice.
2. Incubate mixture overnight at 4°C on a rocking platform.
3. Spin down Sepharose beads (full speed for 1 min in a microfuge) at 4°C, remove the supernatant, and set aside.
4. To wash immunoprecipitates, resuspend beads in 0.5 mL IP buffer, vortex, spin down beads as above, and discarded the supernatant; repeat this washing process three more times.
5. After the last wash, spin once more, and remove all the excess buffer except approx. 10 µL covering the Sepharose gel.
6. Add 20 µL 2X LaemmLi buffer, vortex, boil for 2 min, spin for 2 min in a microcentrifuge, and load 20 µL sample onto a 15% polyacrylamide gel.
7. Run gel for 1 h at 200 V, fix, and process for fluorography as described elsewhere ***(16)***.
8. Mix 1 µL of the remaining IP eluate (**step 6** *above*) with 20 µL 1X LaemmLi buffer, and load onto a separate polyacrylamide gel. This gel is subjected to Western blot analysis with anti-Ras antibody in order to control the efficacy of immunoprecipitation (*see below*).

3.7. Immunoblotting Studies

For immunoblotting studies using crude oocyte extracts, use 1 µL of cell lysate (0.1 oocyte equivalent) with per lane; in all other cases load the indicated amount.

1. Separate proteins by SDS-PAGE, and transfer to nitrocellulose sheets by electroblotting according to the manufacturer's recommendations.
2. Block membranes in blocking buffer overnight at 4°C.
3. Incubate for 2 h at room temperature with 10 mL primary antibody diluted in blocking buffer (anti-MAPK, 0.1 µg/mL; anti-Ras, 0.25 µg/mL; antiphosphotyrosine, 2 µg/mL) on a rocking platform (*see* **Note 4**).
4. Wash membrane for 1 h on a rocking platform with five to six portions of 100 mL washing buffer.
5. Incubate for 2 h at room temperature on a rocking platform with 10 mL secondary antibody diluted 1/1000 in blocking buffer.
6. Wash membrane for 1 h on a rocking platform with five to six portions of 100 mL washing buffer.
7. Detect antibodies bound by enhance chemoluminescence using the ECL kit following the manufacturer's protocol.

3.8. Studies with H-Ras Containing Farnesyl Analogs

1. To 100 µL of 2X PFT buffer in an Eppendorf tube, add 1 nmol of H-Ras, 50 pmol recombinant PFT, and 10 nmol of the respective lipid pyrophosphate, and adjust

the volume to 200 µL with water. Include control reactions where either the lipid pyrophosphate or PFT has been omitted from the reaction mixture.

2. Vortex briefly, and incubate for 4 h in a water bath at 30°C.
3. Stop reactions by placing tubes on ice.
4. To each of the in vitro lipidation reactions, add 25 µL of GTP (100 µ*M* final) followed by 25 µL of 10X nucleotide exchange buffer.
5. Return tubes to the 30°C water bath for 20 min to allow nucleotide exchange.
6. Stop nucleotide exchange by adding 5 µL $MgCl_2$ (20 m*M* final) on ice.
7. Add 12 µL of 2% Octyl-glucoside (0.1% final) to prevent sample loss during subsequent steps.
8. To remove excess GTP and unreacted lipid-pyrophosphates, load 100-µL sample onto a 0.5-mL P-6 Biogel spin column equilibrated with Ras buffer. Spin sample at approx 2000*g* for 2 min in a swing-out rotor in a clinical centrifuge at room temperature. Arrange the spin column such that the eluate is collected directly into the sample chamber of a passivated Microcon-10 unit.
9. Add an additional 20 µL of Ras buffer, and spin out as above.
10. Using a fresh spin column, repeat **steps 8** and **9** with a second 100-µL portion of the reaction mixture. Keep the remainder of the reaction mixture (about 50 µL) for SDS-PAGE analysis (*see* **Note 7**).
11. Concentrate samples at 10,000*g* for 45 min at 4°C in a passivated Microcon-10 unit using a fixed-angle rotor to a final volume of approx. 30–40 µL
12. Recover sample, determine Ras concentration by measuring GTP binding as described ***(11)***, dilute to 10 µM with Ras buffer containing 0.1% octyl-glucoside and flash-freeze 5-µL aliquots in liquid nitrogen; store at –80°C.

Microinjection:

Microinject 50 nL (0.5 pmol) of lipidated H-Ras into oocytes as described above. The microinjection procedure and methods for the analysis of posttranslational processing, oocyte maturation, and kinase activation are identical to those described above, except that buffer supplemented with 0.1% octyl-glucoside is used in control injections.

4. Notes

1. The antibody-based affinity-purification procedure yields H-Ras of high purity (essentially no other protein bands detectable on Coomassie-stained SDS-PAGE). However, we find that the recovery is quite poor, which is most likely owing to the limited binding capacity of the affinity column (most H-Ras is in the flowthrough). To improve the effectiveness of the purification, the initial flowthrough can be reapplied to the regenerated affinity resin. By repeating the purification procedure with this "recycled" lysate, an amount of H-Ras quantitatively and qualitatively comparable to that isolated during the first run can be recovered. The affinity resin is regenerated by washing first with 10 vol of 0.1 m*M* glycine, pH 2.5, followed by equilibration with 10 vol of loading buffer (*see* **ref.** ***12***).

2. Although some microinjection devices offer variable injection volumes, the range of accurately injectable volumes is not wide enough to allow a complete dose–response curve using a single H-Ras stock solution. We therefore prefer to microinject equal volumes of serially diluted stock solutions of H-Ras.
3. It is important to carry out all operations for Triton X-114 phase partitioning requiring a homogeneous solution on ice or at 4°C, and to work at room temperature or higher if detergent phase separation is desired.
4. The primary antibody solutions in blocking buffer can be reused several times if desired. Add sodium azide (0.02% final), and store at 4°C up to 1 mo. For phosphotyrosine immunoblots, include two additional washes of 30 min each after the secondary antibody incubation in order to reduce nonspecific binding.
5. To rule out that the H-Ras preparation used in this study contains any trace amounts of processed H-Ras, H-Ras is expressed in *Escherichia coli* (which lacks the enzymatic machinery for protein prenylation) instead of isolating the nonprocessed protein from the aqueous phase of Triton-phase partitioned Sf9 cells ***(10,11)***.

   It is important to ascertain that the bacterially expressed H-Ras can be stoichiometrically prenylated, since nonprenylated H-Ras can have a dominant-negative effect on H-Ras function in intact cells ***(17)***. Great care has to be taken to avoid C-terminal proteolysis during the isolation of H-Ras from *E. coli* lysates. We found that the most reliable preparations of H-Ras can be obtained if bacterial lysis and all subsequent purification steps are carried out at 4°C in the absence of magnesium and nucleotide in buffer containing 0.1 m*M* EDTA ***(11)***. Although there has been some discussion that Ras proteins can denature irreversibly in the absence of nucleotide and magnesium, the preparations obtained using the described procedure did not reveal a reduction in GTP- binding or biological activity.
6. The washing of the centricon with buffer after spinning out the initial concentrate is optional and has been introduced to optimize the total protein recovery.
7. To verify completion of the lipidation reaction, mix 25 µL of the remaining sample (**Subheading 3.8.**, **step 10**) with 6 µL of 5X LaemmLi buffer, boil for 5 min, and load onto a 15% polyacrylamide minigel (Bio-Rad). Allow to migrate for 75–80 min at 200 V followed by staining the gel with Coomassie. Lipidation is manifested on these gels by an increased apparent electophoretic mobility of the lipidated compared to the nonlipidated from (for illustration, *see* **ref. *11***).

   The increase in electrophoretic mobility for the geranylated form of H-Ras is not always clearly visible. Since all the lipid-pyrophosphates used are tritiated, the stoichiometry of geranylation can be evaluated radiometrically by cutting out the H-Ras band from the stained gel, dissolving the gel slice in 30% H_2O_2, and submitting the sample to scintillation counting ***(18)***.

5. Concluding Remarks

The methodology described in this chapter has been optimized for studies analyzing the contribution of H-Ras palmitoylation and the significance of the H-Ras farnesyl-group for signal transduction and cell-cycle progression in

Xenopus oocytes. Representative examples of experimental data are presented elsewhere ***(11,19)***. The overall experimental approach described here, however, should also prove useful for studies of other lipidated proteins for which a suitable cellular test system exists.

References

1. Herskowitz, I. (1995) Map kinase pathways in yeast: for mating and more. *Cell* **80,** 187–197.
2. Casey, P. J. (1995) Protein lipidation in cell signaling. *Science* **268,** 221–225.
3. Lowry, D. R. and Willumsen, B. M. (1993) *Ann. Rev. Biochem.* **62,** 851–891.
4. Schafer, W. R. and Rine, J.. (1992) Protein prenylation: genes, enzymes, targets and functions. *Ann. Rev. Genet.* **30,** 209–237.
5. Clarke, S. (1992) Protein isoprenylation and methylation at the carboxy-terminal cysteine residue. *Annu. Rev. Biochem.* **61,** 355–386.
6. Hancock, J. F., Magee, A. I., Childs, J. E., and Marshall, C. J. (1989) All ras proteins are polyisoprenylated but only some are palmitoylated. *Cell* **57,** 1167–1177.
7. Bhattacharya, S., Chen, L., Broach, J. R., and Powers, S. (1995) Ras membrane targeting is essential for glucose signaling but not for viability in yeast. *Proc. Natl. Acad. Sci. USA* **92,** 2984–2988.
8. Hancock, J. F., Paterson, H., and Marshall, C. J. (1990) A polybasic domain or palmitoylation is required in addition to the CAAX motif to localize $p21^{ras}$ to the plasma membrane. *Cell* **63,** 133–139.
9. Buss, J. E. and Sefton, B. M. (1986) Direct identification of palmitic acid as the lipid attached to $p21^{ras}$. *Molec. Cell. Biol.* **6,** 116–122.
10. Dudler, T. and Gelb, M. H. (1996) Palmitoylation of Ha-Ras facilitates membrane binding, activation of downstream effectors, and meiotic maturation in *Xenopus* oocytes. *J. Biol. Chem.* **271,** 11,541–11,547.
11. McGeady, P., Kuroda, S., Shimizu, K., Takai, Y., and Gelb, M. H. (1995) The farnesyl group of H-Ras facilitates the activation of a soluble upstream activator of MAP-kinase. *J. Biol. Chem.* **270,** 26,347–26,351.
12. Porfiri, E., Evans, T., Bollag, G., Clark, R., and Hancock, J. F. (1995) Purification of baculovirus expressed recombinant Ras and Rap proteins. *Methods Enzymol.* **255,** 13–21.
13. Celis, J. E. (1994) *Cell Biology: A Laboratory Manual.* Academic, San Diego.
14. Maller, J. L. and Koontz, J. W. (1981) A study of the induction of cell division in amphibian oocytes by insulin. *Dev. Biol.* **85,** 309–316.
15. Birchmeier, C., Broek, D., and Wigler. M. (1985) Ras proteins can induce meiosis in *Xenopus* oocytes. *Cell* **43,** 615–621.
16. Magee, A. I., Wootton, J., and De Bony, J.. 1995. Detecting radiolabeled lipid-modified proteins in polyacrylamide gels. *Methods Enzymol.* **250,** 330–336.
17. Gibbs, J. B., Schaber, M., Schofield, T. L., Scolnick, E. M., and Sigal, I. S. (1989) Xenopus oocyte germinal-vesicle breakdown induced by [Val12]Ras is inhibited by a cytosol localized mutant. *Proc. Natl. Acad. Sci. USA* **86,** 6630–6634.

18. Liu, L., Dudler, T., and Gelb, M. H. (1996) Purification of a Protein Palmitoyltransferase that acts on H-Ras protein and on a C-terminal N-Ras peptide. *J. Biol. Chem.* **271,** 23,269–23,276.
19. Dudler, T. and Gelb, M. H. (1997) Replacement of the H-Ras farnesyl group by lipid analogues: Implications for downstream processing and effector activation in *Xenopus* oocytes. *Biochemistry* **34,** 12,434–12,441.

13

Fluorescence Measurement of Lipid-Binding Affinity and Interbilayer Transfer of Bimane-Labeled Lipidated Peptides

John R. Silvius

1. Introduction

1.1. Background

1.1.1. Theoretical Aspects

The binding of a variety of bimane-labeled lipidated peptides to lipid vesicles substantially enhances the fluorescence intensity of the bimanyl group *(1)*. This property makes it possible to monitor with high precision the binding of bimane-labeled lipidated peptides to lipid vesicles, avoiding complications that can arise in assays requiring physical separation of free and vesicle-bound populations of amphiphilic molecules. The bimane group itself appears to have little effect on the surface-binding affinity of labeled peptides ***(1–3)***.

If a lipidated peptide is present in very low concentrations relative to the lipid phase, so that peptide–peptide interactions at the vesicle surface can be ignored, the equilibrium distribution of the peptide between the aqueous phase and the lipid vesicles can be described in terms of a simple partition coefficient, *Kp*:

$$Kp \equiv x\ peptide\ (bilayer)/xpeptide\ (aqueous) \tag{1}$$

where *xpeptide* represents the mol fraction of peptide in the aqueous or lipidic phase, or in terms of an effective dissociation constant, $Kdeff \equiv 55.5\ M/Kp$. At low concentrations of peptide relative to lipid, the relationship between the lipid concentration and the fraction of vesicle-bound peptide has a simple hyperbolic form:

$$Fraction\ of\ peptide\ bound = \{Lipid\}eff/(Kdeff + \{Lipid\}eff) \tag{2}$$

From: *Methods in Molecular Biology, Vol. 116: Protein Lipidation Protocols*
Edited by: M. H. Gelb © Humana Press Inc., Totowa, NJ

where [Lipid]*eff* represents the total concentration of lipid to which the peptide has access during the partitioning measurement. Since the fluorescence of the vesicle-bound form of the peptide is greater than that of the aqueous form, the fluorescence of a fixed quantity of lipidated peptide will therefore increase in a hyperbolic manner as a function of the concentration of added lipid, as illustrated in **Fig. 1A**:

$$F = Fo + \{\Delta F_{\max}\} \cdot \{lipid\}eff / \{Kdeff + [lipid]eff\} \quad (3)$$

where F is the fluorescence measured at any given lipid concentration, Fo is the fluorescence value that would be observed for the same quantity of peptide in the complete absence of lipid, and (Fo + $\Delta Fmax$) is the fluorescence of the same quantity of peptide at saturating lipid concentrations (when the peptide is entirely vesicle-bound). For peptides that rapidly access the inner as well as the outer surfaces of vesicles (by rapid transbilayer diffusion), the effective lipid concentration [Lipid]*eff* is equal to the total lipid concentration. By contrast, for peptides that do not exhibit rapid transbilayer diffusion (as is the case for most charged peptides), this quantity is equal to the concentration of lipids exposed at the outer surfaces of the lipid vesicles. A method to determine the rates of transbilayer diffusion of bimane-labeled lipidated peptides is described in Chapter 14, and a method to estimate the fraction of surface-exposed lipids in a vesicle preparation is described in **Subheading 3.5**.

1.1.2. Preparation of Bimane-Labeled Lipidated Peptides

Detailed methodologies for the preparation and bimanyl labeling of lipidated peptides are outside the scope of this chapter. Monobromobimane can be used to label selectively the cysteine side chains of peptides under mild conditions ***(4)***. Alternatively, the author has found that peptides can be labeled during Fmoc-based solid-phase synthesis by coupling S-bimanylmercaptoacetic acid as the N-terminal residue. Preparation of S-bimanylmercaptoacetic acid is straightforward ***(5)***; detailed protocols for the synthesis of this compound, and for its coupling to peptides during solid-phase synthesis, can be obtained from the author.

A variety of methods are available for the preparation of lipidated peptides. N-terminal fatty acyl groups can be readily incorporated into peptides in the final coupling cycle of a solid-phase peptide synthesis. Prenylation of peptides is usually accomplished by reaction of a protected peptide with a prenyl halide ***(1,6,7)***, or by coupling a small prenylated peptide fragment to a larger protected peptide ***(8)***. Peptides with no free amino, carboxyl, or hydroxyl groups can be S-acylated using the appropriate acyl chloride; other types of peptides can be S-acylated by incubation with micellar acyl-coenzyme A dispersions, if the peptide is sufficiently hydrophobic and/or basic to interact with the micelles ***(5,9,10)***.

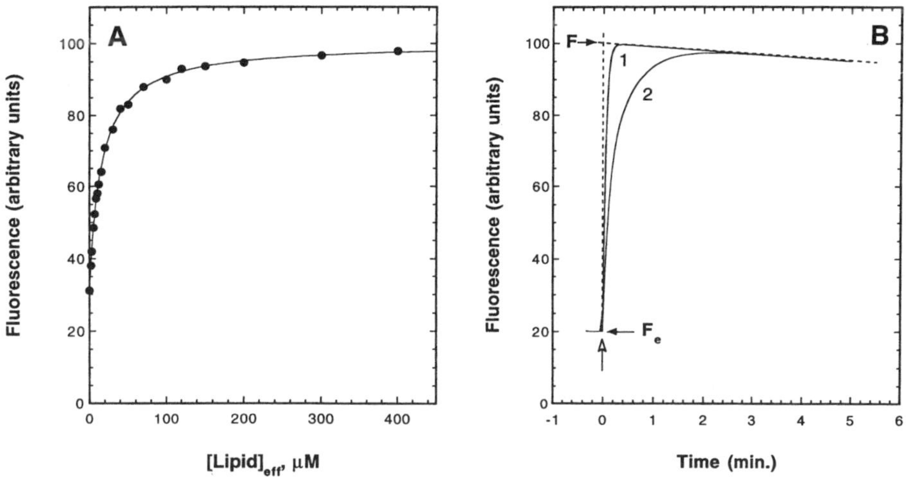

Fig. 1. **(A)** Hypothetical titration of the fluorescence of a lipidated peptide with lipid vesicles, using data acquired as described in **Subheading 3.3.** and as analyzed in **Subheading 3.6.** The data are fitted to **Eq. 3** (solid line) by nonlinear least-squares analysis. **(B)** Hypothetical time courses of fluorescence evolution when a sample of lipidated peptide, incorporated in a small amount of carrier vesicles, is added at time zero to peptide-free vesicles already present in the fluorimeter cuvet. The fluorescence signal caused by light-scattering from the peptide-free vesicles initially present in the cuvet (*Fe*) is determined as indicated. The fluorescence attained upon adding the peptide-containing vesicles is determined by back-extrapolating the slowly decaying plateau to the moment of addition of the peptide-containing vesicles (dashed line). In curve 1, transbilayer diffusion of the lipidated peptide is not a factor in the evolution of the fluorescence response (either because transbilayer diffusion is very fast or because it is very slow on the time scale of the fluorescence measurement). In curve 2, transbilayer flip-flop proceeds at an intermediate rate, so that a few minutes are required for the fluorescence signal to restabilize upon injection of the peptide-loaded vesicles into the cuvet.

1.2. Experimental Strategy

1.2.1. Measurement of Peptide Partitioning into Vesicle Bilayers

In principle, measurements of peptide partitioning into vesicles could be carried out simply by adding increasing amounts of lipid vesicles to a single portion of lipidated peptide in the fluorimeter cuvet, and recording the fluorescence value after each addition. However, in practice, better results are usually obtained by carrying out mutiple runs in which replicate aliquots of peptide, preincubated with small amounts of lipid vesicles as a carrier, are added to different amounts of lipid vesicles in the fluorimeter cuvet.

The experimental procedure for the method just noted is straightforward: for each run, a different concentration of lipid vesicles (without peptide) is first placed in a stirred fluorimeter cuvet, and the baseline signal caused by light-scattering is measured. A small aliquot of preincubated vesicle–peptide mixture is then added, while recording the fluorescence continuously. The final fluorescence plateau value is determined for each of a series of runs, using a range of lipid concentrations that span both the rising and the plateau regions of the partitioning curve, as shown in **Fig. 1A**.

1.2.2. Determination of Total Lipid Concentration

The concentrations of lipid in any extruded vesicle preparation should be measured directly, to correct for possibly variable loss of lipid during the filter-extrusion step. A particularly reliable method is to assay the final vesicle preparation for lipid phosphorus, using a method such as the modified assay of Lowry and Tinsley ***(11)***, described in this chapter.

1.2.3. Determination of Concentration of Peptide-Accessible Lipids

To interpret the results of water–bilayer partitioning measurements like those described here, it is important to determine whether the peptide can diffuse rapidly across lipid bilayers, thus gaining access to all of the vesicle lipids on the time-scale of the partitioning measurements, or, instead exhibits slow transbilayer diffusion, and thus has rapid access only to the lipids present in the outer surfaces of the vesicles. Methods to address this question are described in Chapter 14.

If transbilayer diffusion of the lipidated peptide is slow, the size of the surface-exposed lipid pool is determined by incorporating into the vesicles a modest proportion of an aminophospholipid such as phosphatidylethanolamine (PE), whose surface-exposed fraction can be assayed using the bilayer-impermeant reagent trinitrobenzene-sulfonic acid (TNBS), in a modification of the procedure of Nordlund et al. ***(12)***. In this assay, TNBS is used to convert PE to a chromophoric derivative, either in intact vesicles (assaying the surface-exposed PE fraction only) or in detergent-solubilized vesicles (assaying the total PE pool). It has been shown that PE distributes symmetrically between the external and internal faces of large lipid vesicles like those used here ***(12)***, so that the fraction of PE accessible to TNBS in intact vesicles can be equated to the fraction of total lipids exposed at the vesicle surfaces. This quantity, taken together with the total lipid concentration determined as described above, allows accurate estimation of the concentration of peptide-accessible lipids ([Lipid]*eff*) for partitioning-affinity determinations using peptides that exhibit slow transbilayer flip-flip.

2. Materials

2.1. Preparation of Large Unilamellar Vesicles by Extrusion

1. Phosphatidylcholine purified from egg yolk (Avanti Polar Lipids, Alabaster, AL), stock solution in chloroform stored under nitrogen or argon at –20°C.
2. Phosphatidylethanolamine (Avanti Polar Lipids), prepared by transphosphatidylation from egg phosphatidylcholine, stock solution in chloroform stored under nitrogen or argon at –20°C.
3. 13 × 100-mm Glass tube.
4. Nitrogen source.
5. Vacuum (oil) pump.
6. Vacuum dessicator.
7. Vesicle buffer: 150 m*M* NaCl, 10 m*M* HEPES, 0.1 m*M* EDTA, pH 7.4.
8. Vortex mixer.
9. Parafilm.
10. 50- or 100-mL beaker.
11. Ethanol.
12. Dry ice.
13. A suitable extrusion device, either hand-operated (available from Avestin, Ottawa, Canada, or Avanti Polar Lipids) or cylinder-type (available from Lipex Extruders, Richmond, British Columbia, Canada).
14. 0.1-µm Pore size polycarbonate filters of appropriate diameter to fit the extrusion device.

2.2. Peptide Loading into Vesicles

1. Lipid vesicles, prepared as in **Subheading 2.1**.
2. Lipidated peptide, ≥1 m*M* stock solution in methanol or dimethylformamide.
3. Glass tubes (13 × 100 mm).
4. Aluminum foil.

2.3. Fluorescence Assay of Peptide–Vesicle Binding

1. 13 × 100 mm glass tubes.
2. Fluorimeter with stirred, thermostatted cuvet holder.
3. Vesicle buffer: 150 m*M* NaCl, 10 m*M* HEPES, 0.1 m*M* EDTA, pH 7.4.
4. Peptide-loaded and peptide-free vesicles, prepared as in **Subheadings 2.1.** and **2.2.** above.

2.4. Determination of Lipid Concentration

1. Glass-distilled water.
2. 16 × 125-mm screw-capped Pyrex tubes, washed first with 2% acetic acid in glass-distilled water, then with methanol, and dried.
3. 10 m*M* potassium or sodium phosphate standard, stored at –20°C.
4. 70–72% perchloric acid, reagent grade.
5. Concentrated nitric acid, reagent grade.
6. Heating block, with wells for 16-mm tubes, and capable of heating to at least 130°C.

7. *n*-Butyl acetate, reagent grade.
8. Ammonium molybdate reagent: In a fume hood, dissolve 14.6 g ammonium molybdate (reagent grade) in 50 mL reagent grade HCl, then add 200 mL glass-distilled water.
9. Clinical centrifuge (optional).
10. Spectrophotometer.

2.5. Determination of Fraction of Surface-Exposed Lipid

1. Vesicle suspension, prepared as in **Subheading 2.1**.
2. 13 × 100-mm tubes.
3. Vesicle buffer: 150 m*M* NaCl, 10 m*M* HEPES, 0.1 m*M* EDTA, pH 7.4.
4. Glass-distilled water.
5. Vortex mixer.
6. 1.6% (v/v) Triton X-100 in 0.8 *M* $NaHCO_3$, pH 8.5.
7. 0.8 *M* $NaHCO_3$, pH 8.5.
8. 1.5% (w/v) Trinitrobenzenesulfonic acid (TNBS), prepared by diluting the commercially available 5% solution (Pierce) with glass-distilled water.
9. 0.4% (v/v) Triton X-100 in 1.5 N HCl.
10. 1.2% (v/v) Triton X-100 in 1.5 N HCl.
11. Spectrophotometer.

3. Methods

3.1. Preparation of Large Unilamellar Vesicles by Extrusion

1. Dispense 18 µmol phosphatidylcholine and 2 µmol phosphatidylethanolamine into a 13 × 100-mm disposable tube (*see* **Note 1**). Dry down, first under nitrogen, and then in a desiccator under high vacuum (oil pump), for at least 6 h.
2. Suspend the lipid by vortexing for at least 1 min in 1 mL vesicle buffer.
3. Cover the tube with parafilm and freeze it in a small beaker containing an ethanol–dry-ice bath.
4. After the sample is completely frozen, allow it to thaw completely, and vortex.
5. Repeat **steps 3** and **4** a total of five times.
6. After the final thawing, extrude the sample through 100-nm pore size polycarbonate filters, according to the manufacturer's instructions. The nominal concentration of lipid in the extruded sample is 20 m*M*; the true concentration is determined later, as described in **Subheading 3.4**.

3.2. Peptide Loading into Vesicles

1. Mix 15 µL of the above vesicle suspension and 285 µL vesicle buffer in a 13 × 100-mm disposable glass tube.
2. Inject 3 nmol peptide (maximum 3 µL vol), from a methanol or dimethyl-formamide stock solution, into the vesicle sample; gently mix the sample, flush the tube with nitrogen or argon, seal with parafilm, wrap in foil, and incubate for 15 min at 37°C.
3. Store the sample at room temperature in a foil-jacketed tube, to exclude light.

3.3. Fluorescence Assay of Peptide–Vesicle Binding in Dilute Solution

1. Into a stirred fluorimeter cell containing 3 mL vesicle buffer, inject 10 nmol (= 10 μL) peptide-loaded vesicles, while continuously recording the fluorescence. The fluorescence should be recorded continuously (on a strip-chart recorder or computer file), until it has stabilized, or until a very slow, essentially linear decrease in fluorescence (caused by adsorptive loss of peptide onto the walls of the cuvet) is observed, as shown in **Fig. 1B**.
2. Repeat **step 1** with the following modification: Add first to the cuvet an accurately measured quantity of lipid vesicles (without peptide), and allow the fluorescence reading to stabilize, then add 10 nmol of peptide-loaded vesicles, and continue recording the fluorescence as above.
3. Repeat **step 2** a minimum of 12–15 times, each time initially adding to the cuvet different concentrations of peptide-free vesicles, followed by 10 nmol of peptide-loaded vesicles. The range of added lipid concentrations is chosen to span both the rising and the plateau phases of the hyperbolic curve shown in **Fig. 1A** (*see* **Note 2**).
4. Measure the fluorescence of 10 nmol peptide-free vesicles in 3 mL buffer. This measurement allows correction (usually minor) for light scattering from the carrier vesicles (F_c) in the calculations described in **Subheading 3.6.**
5. If the kinetics of intervesicle peptide exchange are also to be examined (*see* Chapter 14), measure the fluorescence of 10 nmol of the peptide-loaded vesicles in 3 mL methanol. This quantity is used in the calculations described in **Subheading 3.4.** of Chapter 8.

3.4. Determination of Lipid Concentration

The following protocol is a modification of the procedure of Lowry and Tinsley *(11)*.

1. Mix 20 μL of the concentrated vesicle stock prepared in **Subheading 3.2.** (nominal concentration 20 m*M*) with 20 μL vesicle buffer.
2. Into three 16 × 100-mm screw-capped tubes (sample) dispense 10 μL of this diluted vesicle suspension.
3. Into a second set of three tubes (blank) dispense 10 μL of buffer.
4. Into a third set of three tubes (standard) dispense 10 μL of 10 m*M* phosphate solution, plus 10 μL of buffer.
5. To each digestion tube add, in turn, 0.5 mL perchloric acid and 2 drops nitric acid. Incubate the tubes in a fume hood for 2 h in a prewarmed heating block at 130°C, adding 1 drop more nitric acid after the first half-hour of digestion.
6. At the end of the digestion period, cool the tubes and add in succession 7.5 mL glass-distilled water, 0.5 mL ammonium molybdate reagent, and 2.5 mL butyl acetate. Cap the tubes and shake vigorously for a minimum of 20 s. Allow the layers to separate and the upper phase to clarify by letting stand for 30 min, or by centrifuging at low speed in a clinical centrifuge for 2 min.

7. From each tube, carefully withdraw approx 1.5 mL of the upper phase, and read the absorbance at 310 nm.
8. The concentration of lipid in the original vesicle stock suspension is determined as:

$$\{Lipid\}stock = \{Aavg[sample] - Aavg[blank]\}/ \{Aavg[standard] - Aavg[blank]\} \cdot \{20\ mM\} \quad (4)$$

3.5. Determination of Fraction of Surface-Exposed Lipid

The assay described below is a modification of the procedure of Nordlund et al. ***(12)***.

1. Accurately dispense 50 µL of the original vesicle suspension (*see* **Note 3**) into each of four 13 × 100-mm disposable glass tubes; label two of these "Total" and the other two "Surface."
2. Dispense 50 µL vesicle buffer into four other glass tubes; label two of these "Blank (total)" and the other two "Blank (surface)."
3. To each tube, add 550 µL distilled water, and vortex.
4. To the tubes marked "Total" and "Blank (total)," add 200 µL of 1.6% Triton X-100 in 0.8 *M* $NaHCO_3$, pH 8.5, and vortex. The samples marked "Total" should clarify during this step.
5. To the tubes marked "Surface" and "Blank (surface)," add 200 µL of 0.8 *M* $NaHCO_3$ and vortex.
6. To each of the tubes, add 20 µL 1.5% TNBS, mix well, and incubate for exactly 10 min at room temperature. The time of addition of the TNBS reagent to each tube should be staggered, so that the appropriate acidic quenching solution can be added to each tube exactly 10 min after addition of TNBS.
7. Precisely 10 min after the addition of the TNBS reagent, to each tube marked "Total" and "Blank (total)" add 0.4 mL of 0.4% Triton X-100 in 1.5 N HCl; mix well. To each tube marked "Surface" and "Blank (surface)," add 0.4 mL of 1.2% Triton X-100 in 1.5 N HCl and mix.
8. Within 30 min of the addition of the latter reagent read the sample absorbances at 410 nm.
9. The fraction of surface-exposed lipid (which for large vesicles is equal to the fraction of surface-exposed phosphatidylethanolamine ***[12]***) is determined by the equation

$$Fraction\ of\ surface\ PE = \{Aavg[Surface] - Aavg[Blank(surface)]\}/ \{Aavg[Total] - Aavg[Blank(total)]\} \quad (5)$$

3.6. Data Analysis

1. Two representative types of fluorescence traces that may be observed in individual peptide-binding runs are shown schematically in **Fig. 1B**. The first type of trace is observed when transbilayer diffusion of the peptide is either very rapid or very slow on the time-scale of the fluorescence measurement. The second type of

trace is observed when the rate of transbilayer diffusion is intermediate, so that complete equilibration of peptide molecules between the outer and inner monolayers of the lipid vesicles may require incubation in the cuvet for a few minutes.

2. Referring to the fluorescence time-courses shown in **Fig. 1B**, for each trace, determine the fluorescence (F) extrapolated back to time zero, as indicated. This extrapolation corrects for slow loss of the fluorescent species caused by adsorption on the walls of the cuvet.
3. For each trace, determine the fluorescence value (F_e) ascribable to light-scattering from the peptide-free vesicles added first.
4. Record the fluorescence correction for light-scattering from the carrier vesicles **(Fc)** determined, as described in **Subheading 3.3., step 4**. This correction may be negligible.
5. Set up a worksheet for the runs as shown below:

Volume Peptide-free Vesicles (V_e)	nmol Peptide-free Vesicles (L_e)	F	F_e	F_c	ΔF_{cor}	$[L]_{eff}$

The relevant calculations for the last two columns are carried out as follows:

$$\Delta Fcor = \{F - [Fe + Fc]\} \cdot \{Vo + Vc + Ve\}/\{Vo + Vc\} \tag{6}$$

$$Leff = \{Lc + Le\}/\{Vo + Ve + Vc\} \cdot \gamma \tag{7}$$

where Vo is the volume of buffer present in the cuvet before addition of lipid in each run, Lc = 10 nmol is the amount of peptide-loaded vesicles added, and the other quantities are as defined above. The parameter γ equals 1 for conjugates that exhibit rapid transbilayer diffusion; for conjugates exhibiting slow transbilayer diffusion on this-time scale, γ equals the fraction of lipid exposed at the vesicles' outer surfaces (determined as described in **Subheading 3.5.**). The rate of peptide transbilayer diffusion is determined as described in Chapter 14. Plot the corrected fluorescence value $\Delta Fcor$ vs the effective lipid concentration $Leff$ and fit to **Eq. 3** above, using a standard nonlinear regression analysis with the values Fo, $\Delta Fmax$, and $Kdeff$ as adjustable parameters.

4. Notes

1. Phospholipid stock solutions in organic solvents should be dispensed using glass pipets or microliter syringes, rather than micropipetors; the latter are inaccurate for dispensing volatile solvents, and disposable plastic tips may leach contaminants into the organic solution.
2. Occasionally, a peptide may show only a small enhancement of fluorescence on vesicle binding, depending on the site of bimane labeling. In such cases, the inclusion of 1–2% NBD-PE (available from Avanti Polar Lipids, or from Molecular Probes, Junction City, OR) in the lipid vesicles allows the binding of the peptides to be monitored by quenching based on resonance energy transfer *(13)*.
3. Vesicles with lipid compositions other than that suggested in **Subheading 3.1.** can be used. However, for success of the TNBS procedure to assay the fraction of surface-exposed vesicle lipids (**Subheading 3.5.**) requires that the vesicles contain at

least 5%, and at most approx 25%, aminophospholipids (phosphatidylethanolamine and/or phosphatidylserine). In this case, the amount of lipid vesicles used in the TNBS assay should be adjusted to give ca. 0.1 µmol of aminophospholipid per sample.

References

1. Silvius, J. R. and l'Heureux, F. (1994) Fluorimetric evaluation of the affinities of isoprenylated peptides for lipid bilayers. *Biochemistry* **33,** 3014–3022.
2. Skerjanc, I., Shore, G. C., and Silvius, J. R. (1987) The interaction of a synthetic mitochondrial signal peptide with lipid membranes is independent of transbilayer potential. *EMBO J.* **6,** 3117–3123.
3. Leventis, R., Juel, G., Knudsen, J. K., and Silvius, J. R. (1997) Acyl-CoA binding proteins inhibit the nonenzymic S-acylation of cysteinyl-containing peptide sequences by long-chain acyl-CoAs. *Biochemistry* **36,** 5546–5553.
4. Kajiwara, K., Kunazaki, T., Sato, E., Kanaoka, Y., and Ishii, S. (1991) Application of bimane-peptide substrates to spectrofluorimetric assays of metalloendo-peptidases. *Biochem. J.* **110,** 345–349.
5. Shahinian, S. and Silvius, J. R. (1995) Doubly lipid-modified protein sequences exhibit long-lived anchorage to lipid bilayer membranes. *Biochemistry* **34,** 3813–3822.
6. Xue, C.-B., Becker, J. M., and Naider, F. (1992) Efficient regioselective isoprenylation of peptides in acidic aqueous solution using zinc acetate as catalyst. *Tetrahedron Lett.* **33,** 1435–1438.
7. Liu. L., Jang, G.-F., Farnsworth, C. C., Yokohama, K., Glomset, J. A., and Gelb, M. H. (1995) Synthetic prenylated peptides: studying prenyl protein-specific endoprotease and other aspects of protein prenylation. *Methods Enzymol.* **250,** 189–206.
8. Xue, C.-B., Becker, J. M., and Naider, F. (1991) Synthesis of S-alkyl and C-terminal analogues of the Saccharomyces cerevisiae **a**-factor. *Int. J. Peptide Protein Res.* **37,** 476–486.
9. Silvius, J. R. and Quesnel, S. (1994) Cysteine-containing peptide sequences exhibit facile uncatalyzed transacylation and acyl-CoA-dependent acylation at the lipid bilayer interface. *Biochemistry* **33,** 13,340–13,348.
10. Bharadwaj, M. and Bizzozero, O. (1995) Myelin P_0 glycoprotein and a synthetic peptide containing the palmitoylation site are both autoacylated. *J. Neurochem.* **65,** 1805–1815.
11. Lowry, R. J. and Tinsley, I. J. (1974) A simple and sensitive assay for lipid phosphorus. *Lipids* **9,** 941–942.
12. Nordlund, J. R., Schmidt, C. F., Dicken, S. N., and Thompson, T. E. (1981) Transbilayer distribution of phosphatidylethanolamine in large and small unilamellar vesicles. *Biochemistry* **20,** 3237–3241.
13. Silvius, J. R., Leventis, R., Brown, P. M., and Zuckermann, M. J. (1987) Novel fluorescent phospholipids for assays of lipid mixing between membranes. *Biochemistry* **26,** 4279–4287.

14

Determination of the Kinetics of Intervesicle Transfer and Transbilayer Diffusion of Bimane-Labeled Lipidated Peptides

John R. Silvius

1. Introduction

1.1. Background

When a bimane-labeled peptide is bound to donor vesicles incorporating small amounts of a nonexchangeable energy-transfer acceptor such as DABS-PC *(1)* or NBD-PE, the fluorescence is strongly quenched. If a population of acceptor vesicles containing neither peptide nor quencher is subsequently added in large excess to the peptide–donor mixture, the redistribution of the peptide between the donor and acceptor populations can be monitored through the resultant time-dependent enhancement of fluorescence.

1.2. Experimental Strategy

The most informative variation of the approach just mentioned uses donor vesicles into which the lipidated peptide has been preincorporated during vesicle preparation. In this case, with peptide molecules initially present at both the inner and outer surfaces of the donor vesicles, the complete time-course of peptide equilibration between donor and acceptor vesicles will reflect the kinetics of both inter- and transbilayer diffusion of the peptide. Three possible outcomes of such experiments are illustrated schematically in **Fig. 1**, and are discussed below.

If transbilayer diffusion of the peptide is very fast, as illustrated in curve 1 of **Fig. 1**, upon addition of the acceptor vesicles the fluorescence shows a simple exponential increase to a plateau reflecting complete equilibration of the peptide between the donor and acceptor vesicles. If, by contrast, transbilayer

From: *Methods in Molecular Biology, Vol. 116: Protein Lipidation Protocols*
Edited by: M. H. Gelb © Humana Press Inc., Totowa, NJ

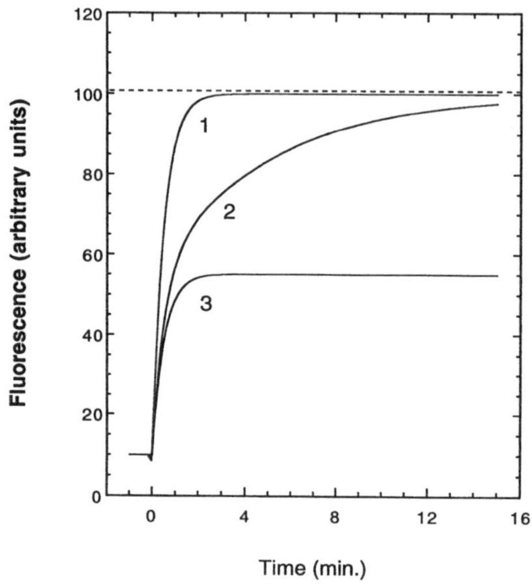

Fig. 1. Hypothetical time-courses of intervesicle redistribution of a lipidated peptide when a small portion of lipid–peptide–quencher donor vesicles is mixed at time zero with a ninefold excess of acceptor vesicles containing neither peptide nor quencher. The dashed line represents the fluorescence value expected upon complete equilibration of the lipidated peptide between the quencher-containing donor and the quencher-free acceptor vesicles. Curve 1 represents rapid redistribution of the peptide between the inner and outer surfaces of the vesicles upon addition of the acceptor vesicles; curve 3 represents rapid redistribution of the peptide only between the outer surfaces of the vesicles. Curve 2 is biphasic, but, at longer times, approaches the fluorescence expected upon complete redistribution of the lipidated peptide, indicating that flip-flop of the peptide from the inner to the outer monolayer of the donor vesicles proceeds only gradually over the time-scale of the fluorescence measurement.

diffusion of the peptide is very slow on the time-scale of the transfer measurements, the fluorescence time-course will again be a simple exponential, but will plateau at a level reflecting the redistribution only of the peptide molecules that were initially present at the outer surfaces of donor vesicles (curve 3 of **Fig. 1**). Finally, if the peptide exhibits an intermediate rate of transbilayer diffusion, the time course of fluorescence enhancement will be nonexponential, but will eventually plateau at a level representing complete equilibration of peptide between the donor and acceptor vesicles (curve 2).

Transfer measurements of this type are simplest to analyze when the lipid concentration is sufficient to ensure that the peptide is nearly all (>90%) bound to the donor vesicles at the beginning of the run. The protocols given below assume that this is the case for the donor lipid concentration specified (20 μ*M*), but this assumption can be evaluated using the results of the partitioning measurements described in Chapter 13, and, if necessary, the concentration of donor vesicles in the experimental protocols below can be adjusted to ensure that this condition is fulfilled.

2. Materials

2.1. Preparation of Peptide-Loaded Donor Vesicles

1. Egg yolk phosphatidylcholine (Avanti Polar Lipids, Alabaster, AL), stock solution in chloroform stored under argon or nitrogen at –20°C.
2. NBD-phosphatidylamine (Avanti Polar Lipids, from egg or dipalmitoyl PE, labeled on the ethanolamine group), stock solution in chloroform, stored under argon or nitrogen at –20°C.
3. Lipidated peptide stock in methanol.
4. Aluminum foil.

5–16. Other materials for preparation of extruded vesicles: *See* **items 3–14** from list in **Subheading 2.1.** of Chapter 13.

2.2. Preparation of Acceptor Vesicles

1. Egg yolk phosphatidylcholine, stock solution in chloroform.

2–13. Other materials for preparation of extruded vesicles: *See* **items 3–14** from list in **Subheading 2.1.** of Chapter 13.

2.3. Measurement of Kinetics of Peptide Transfer Between Vesicles

1. Peptide-loaded donor vesicles, prepared as in **Subheading 2.1**.
2. Acceptor vesicles, prepared as in **Subheading 2.2**.
3. Vesicle buffer: 150 m*M* NaCl, 10 m*M* HEPES, 0.1 m*M* EDTA, pH 7.4.
4. Spectrofluorimeter with thermostatted and stirred cuvet holder.

3. Methods

3.1. Preparation of Peptide-Loaded Donor Vesicles

1. In a 13 × 100-mm disposable glass tube mix 5 μmol phosphatidylcholine, 100 nmol NBD-PE, and 5 nmol lipidated peptide. Remove the solvents under a gentle stream of nitrogen with warming, loosely wrap the tube in foil, and dry under high vacuum (oil pump) for a minimum of 6 h.
2. Suspend the lipid by vortexing in 0.5 mL of vesicle buffer. Continue vortexing for at least 1 min.
3. Cover the tube with parafilm and freeze it in a small beaker containing an ethanol–dry-ice bath. Thaw the sample completely, and again vortex it.
4. Repeat **step 3** a total of five times.
5. After the final thawing, extrude the sample through 100-nm pore size polycarbonate filters according to the manufacturer's instructions.

6. Keep the peptide-loaded vesicles at room temperature in a foil-jacketed tube to exclude light.

3.2. Preparation of Acceptor Vesicles

1. In a 13 × 100-mm disposable glass tube, dry down 5 μmol phosphatidylcholine, first under a stream of nitrogen, then under high vacuum for a minimum of 6 h.
2. Rehydrate the dried lipid by vortexing in 1 mL of vesicle buffer for at least 1 min. Freeze–thaw 5× as described in **Subheading 3.1., steps 3** and **4**.
3. Extrude the vesicles through 100-nm pore-size polycarbonate filters, as described in **Subheading 3.1., step 5**.

3.3. Measurement of Kinetics of Peptide Transfer Between Vesicles

1. To 3 mL of vesicle buffer in a stirred fluorimeter cell, add 60 nmol of peptide-loaded donor vesicles, recording the fluorescence continuously over at least 1 min to establish a baseline.
2. Add 540 nmol of acceptor vesicles to the cuvet and record the time-course of the subsequent fluorescence change. Record the fluorescence continuously for 10–15 min, or until the fluorescence signal has clearly reached a plateau or stabilized to a linearly decreasing signal (*see* **Note 1**).
3. Inject 60 nmol of donor vesicles into methanol in the fluorimeter cuvet, while recording the fluorescence continuously. The resultant value (minus the methanol blank reading) will be needed for the calculations described in the next subheading.

3.4. Data Interpretation

1. To interpret the transfer data, it is necessary to estimate the fraction of peptide that has transferred between vesicles when the fluorescence has reached a plateau. To calculate this value, the following data are required:

    a. *F(donor)*, the fluorescence (minus blank) in 3 mL buffer of 60 nmol peptide-loaded donor vesicles.
    b. *FM(donor)*, the fluorescence (minus blank) in 3 mL methanol of 60 nmol peptide-loaded donor vesicles.
    c. *F(acceptor)*, the fluorescence (minus blank) in 3 mL buffer of a sample of peptide (0.1 nmol), which is essentially 100% bound to acceptor vesicles, measured as the plateau fluorescence in a binding curve, determined as described in **Subheading 3.3.** of Chapter 13; and
    d. *FM(acceptor)*, the fluorescence (minus blank) in 3 mL methanol of 0.1 nmol peptide, measured as described in **Subheading 3.3., step 4** in Chapter 13.

    If the latter two values are not available from a previous vesicle-partitioning experiment, they may be estimated by incorporating the lipidated peptide (≥0.4 nmol) into a sample of acceptor vesicles, and injecting replicate portions into 3 mL buffer or 3 mL methanol, respectively. In this case, however, the concentration of acceptor vesicles must be chosen, so that upon dilution into the cuvet, essentially 100% of the lipidated peptide remains bound to the vesicles.

Having determined the above values, and assuming a donor: acceptor ratio of 1:9 as specified in the above protocol, the fluorescence value expected upon complete equilibration of the peptide between the donor and acceptor vesicles is estimated as follows:

$$Feq = 0.9 \cdot \{FM[donor] \cdot F\,[acceptor]\}/\{FM[acceptor]\} + 0.1 \cdot F\{donor\}$$

2. If the fluorescence time course obtained in **Subheading 3.3.** reaches a plateau value closely approximating the value of *Feq* determined from the equation, it can be concluded that the peptide redistributes between the inner and outer monolayers of the vesicles (by rapid trans- as well as interbilayer diffusion) on the time-scale of the fluorescence measurements. If, instead, the fluorescence plateaus at a value no more than one-half the difference (*Feq – F[donor]*), the peptide molecules are equilibrating only between the outer surfaces of the vesicles, indicating that transbilayer diffusion (flip-flop) of the lipidated peptide is slow (*see* **Note 2**). In the first case, the effective lipid concentration in the partitioning analysis described in Chapter 13 equals the total lipid concentration. In the latter case, the effective lipid concentration in the vesicle-partitioning measurements is the concentration of surface-exposed lipid.

4. Notes

1. For a discussion of a modification of this method to determine the rate constant of interbilayer diffusion of lipidated peptides with very slow rates of intervesicle transfer, *see* **ref *1***.
2. In cases in which transbilayer diffusion of the lipidated peptide is substantially slower than its interbilayer exchange, an appropriate quantitative analysis of the time-course of peptide redistribution upon addition of acceptor vesicles can allow determination of the rate constant for transbilayer flip-flop of the peptide, as described elsewhere for fluorescent fatty acids ***(2)***.
3. In cases in which transbilayer flip-flop of a peptide is determined to be neither complete nor negligible over 15 min, the incubation of peptide with carrier vesicles used for partitioning measurements (Chapter 13, **Subheading 3.2.**, **step 3**) can be shortened to as little as 1–2 min, and the peptide-loaded vesicles can subsequently be kept at 0°C.

References

1. Shahinian, S. and Silvius, J. R. (1995) Doubly lipid-modified protein sequences exhibit long-lived anchorage to lipid bilayer membranes. *Biochemistry* **34,** 3813–3822.
2. Storch, J. and Kleinfeld, A. M. (1986) Transfer of long chain fluorescent fatty acids between unilamellar vesicles. *Biochemistry* **25,** 1717–1726.

15

Preparation and Assay of Myristoyl-CoA:Protein *N*-Myristoyltransferase

Rajala V. S. Raju and Rajendra K. Sharma

1. Introduction

Myristoyl-CoA:protein *N*-myristoyltransferase (NMT) catalyzes the cotranslational transfer of myristate from myristoyl-CoA to the amino-terminal glycine residue of a number of cellular, viral, fungal, and oncoproteins ***(1–5)***. These proteins include the catalytic subunit of cAMP-dependent protein kinase, various tyrosine kinases (including $pp60^{src}$, $pp60^{yes}$, $pp56^{lck}$, $pp59^{fyn/syn}$, and cAbl), the β-subunit of calmodulin-dependent protein phosphatase (calcineurin), the myristoylated alanine rich C kinase substrate, the α-subunit of several G-proteins, and several ARF proteins involved in ADP ribosylation ***(1–5)***. A significant problem with the purification and characterization of NMT concerns the available assay procedures. The wide availability and use of peptides allow for convenient substrates. However, the existing procedures to isolate and analyze the reaction products of myristoylation, such as reverse-phase HPLC, are expensive and time-consuming ***(6,7)***. Ion-exchange column chromatography, differential solubilization, adsorption of myristoyl-CoA to acidic alumina, and ion-exchange exclusion of [^{3}H]myristoyl peptides have been utilized as assay methods are faster and more convenient than reverse-phase HPLC ***(8–10)***. However, these procedures are still limited with respect to the number of assays that can conveniently be performed. Here we described a new rapid, reliable, and inexpensive NMT assay based on the binding of [^{3}H]myristoyl-peptide to a P81 phosphocellulose paper matrix ***(11)***.

Myristoytransferase assay is based on the binding of the [^{3}H]myristoyl-peptide to a P81 phosphocellulose paper matrix.

From: *Methods in Molecular Biology, Vol. 116: Protein Lipidation Protocols*
Edited by: M. H. Gelb © Humana Press Inc., Totowa, NJ

$$\text{[}^3\text{H]myristic acid + ATP} \xrightarrow{Mg^{2+}} \text{[}^3\text{H]myristoyl-AMP + PPi} \quad (1)$$

$$\text{[}^3\text{H]myristoyl-AMP+ CoASH} \xrightarrow{\text{Acyl-CoA Synthetase}} \text{[}^3\text{H]myristoyl-CoA + AMP} \quad (2)$$

$$\text{[}^3\text{H]myristoyl-CoA + Peptide} \xrightarrow{\text{NMT}} \text{[}^3\text{H]myristoyl-peptide + CoASH} \quad (3)$$

An alternative spectrophotometric assay method is described below, which does not require the radioisotope and positive charged peptides. The assay is based on the catalytic conversion of the acyl thioester substrate to free thiol as the basis for the assay of NMT ***(12)***.

$$\text{Peptide + myristoyl-CoASH} \xrightarrow{\text{NMT}} \text{myristoyl-peptide + CoASH} \quad (4)$$

Several groups have observed inhibitory activities toward NMT in a variety of cell homogenates ***(13–18)***. A heat-stable, membrane-associated bovine brain 71-kDa NMT inhibitor protein (NIP_{71}) has been purified and characterized ***(18)***. This inhibitor protein has been suggested to be the regulator of many mammalian NMT activities in vivo.

Antibodies are powerful and versatile tools for immunological approaches in biomedical research. Among other application, antibodies have been used to develop the enzyme immunoassay (EIA), to purify proteins existing in native quantities in the cell, to probe structure–function relationship in proteins, and to study the function of membrane proteins in their native environment. Recently, we have developed the EIA procedure for quantitating NMT. Bovine spleen is chosen for purification, because it is available in large quantities. In addition, $pp60^{src}$, a myristoylated oncogene, is present in abundant quantities in the spleen ***(19)***.

Little is known about the control and regulation of NMT. Our studies indicate ***(20)*** that low concentrations of CoA triggered the demyristoylation reaction, whereas higher concentrations of CoA favored myristoylation. In this chapter, we described the methods for NMT activity, inhibitory activity, NMT protein expression, purification of NMT from bovine spleen, and demyristoylation of peptide substrates.

2. Materials

2.1. Tissue Material

Fresh tissues were obtained from either a local slaughterhouse or tissue biopsies from the operating room, and transferred to the laboratory in packed ice.

2.2. Reagents and Column Material

1. [1-^{14}C]myristoyl-CoA (Amersham; 54.7 mCi/mmol).
2. [9,10-^{3}H]myristic acid (Dupont; 39.3 Ci/mmol).
3. SP-Sepharose fast-flow matrix (Pharmacia).
4. Phenyl-Sepharose CL-4B matrix (Pharmacia).
5. DEAE-Sepharose CL-6B matrix (Pharmacia).
6. Superose 12 (HR/30) fast protein liquid chromatrography column (Pharmacia).
7. Activated CH-Sepharose-4B (Pharmacia).
8. Freund's complete adjuvant (Sigma).
9. Freund's incomplete adjuvant (Sigma).
10. NMT–peptide bovine serum albumin (BSA) cojugate.
11. NMT–peptide keyhole lympet hemocyanin (KLH).
12. Peroxidase-conjugated goat antirabbit IgG (Bio-Rad).
13. Psudomonas acyl–CoA synthetase (Sigma).
14. Adenosine triphosphate (Sigma).
15. LiCoA (Sigma).

2.3. Special Laboratory Tools and Materials

1. Waring blender (Bachofer).
2. Air dryer.
3. Microcentrifuge tubes (1.5 mL).
4. P81 Phosphocellulose paper (Whatman).
5. Washing apparatus.
6. Water bath, 30°C.
7. Tissue-culture microtiter plates (Costar Corporation, Cambridge, MA).
8. ELISA recorder.
9. Liquid sintillation counter (Beckman).
10. Vortex mixer.
11. Dialysis bag, 6800 mol-wt cutoff.
12. Amicon concentration unit.
13. Nitrogen cylinder.

2.4. Buffers and Other Solutions

1. Myristoyl-CoA generation buffer: 38.6 m*M* Tris-HCl buffer, pH 7.4, containing 1.93 m*M* dithiothreitol, 19.3 m*M* $MgCl_2$, and 1.93 m*M* EGTA.
2. NMT assay buffer: 0.26 M Tris-HCl, pH 7.4, containing 3.25 m*M* EGTA, 2.92 m*M* EDTA, and 29.25 m*M* 2-mercaptoethanol.
3. Washing buffer: Add 100 mL of cold (4°C) 1 *M* Tris-HCl, pH 8.0 to 4 L of room temperature distilled water. Mix. There is no need to adjust the pH. It will be 7.3.
4. Buffer A: 100 m*M* Tris-HCl, pH 7.4, containing 1 m*M* EGTA, 1 m*M* EDTA, 10 m*M* 2- mercaptoethanol, 1 m*M* benzamidine, 0.2 m*M* phenylmethylsulfonyl fluoride (PMSF), 0.7 µg/mL leupeptin. The protease inhibitors are freshly prepared and added to the buffer.

5. Buffer B: 50 m*M* potassium phosphate, pH 7.0, containing 0.1 m*M* EGTA, 10 m*M* 2- mercaptoethanol, 1 m*M* benzamidine, 0.2 m*M* PMSF, 0.7 μg/mL leupeptin.
6. Buffer C: 25 m*M* potassium phosphate, pH 7.0, containing 0.1 m*M* EGTA, 10 m*M* 2- mercaptoethanol, 1 m*M* benzamidine, 0.2 m*M* PMSF, 0.7 μg/mL leupeptin.
7. Buffer D: 10 m*M* Tris-HCl, pH 7.4, containing 0.1 m*M* EGTA, 10 m*M* 2-mercaptoethanol, 10% (w/v) sucrose,1 m*M* benzamidine, 0.2 m*M* PMSF, 0.7 μg/mL leupeptin.
8. Psudomonas acyl-CoA synthetase (1 U/mL) (Sigma).
9. Adenosine triphosphate (ATP, 50 m*M*).
10. LiCoA (20 m*M*).
11. Triton X-100 (20%).
12. Peptide (5 m*M*).
13. Myristoyl-CoA (1 m*M*).
14. Palmitoyl-CoA (1 m*M*).
15. Stearoyl-CoA (1 m*M*).
16. Coenzyme A (1 m*M*).
17. *N*-5,5'Dithiobis (2-nitrobenzoic acid) (DTNB, 5 m*M*).
18. Glacial acetic acid.
19. Urea (8 *M*).
20. NaCl (0.9%).
21. Sodium carbonate buffer, pH 9.6 (0.05 M).
22. Phosphate-buffered saline, pH 7.4 (PBS, 0.01 *M*).
23. Tris-HCl buffer, pH 8.0 (25 m*M*).
24. Tris-HCl buffer, pH 7.4 (0.1 *M*).
25. Tris-HCl buffer, pH 8.0 (0.1 *M*).
26. Tris-HCl buffer, pH 8.0 (1 *M*).
27. Tris-HCl buffer, PH 7.0 (20 m*M*).
28. Sodium acetate buffer, pH 4.0 (0.1 *M*).
29. Sodium acetate buffer, pH 7.4 (50 m*M*).
30. Glycine, pH 2.5 (100 m*M*).
31. Polyethylene glycol (30%).
32. Glycerol (50%).
33. Sodium azide (0.02%).
34. BSA Fraction V (0.5%) (Sigma).
35. Skimmed milk power (5%).
36. 4-Chloro-1-naphthol.
37. Hydrogen peroxide.

3. Methods

3.1. NMT Assay

3.1.1. Peptide Synthesis

1. Peptides should have a positive charge at carboxyl-terminal end.
2. The following peptides are synthesized by solid-phase protocol developed by Merrifield ***(21)***.

3. Gly-Asn-Ala-Ala-Ala-Ala-Lys$^+$-Lys$^+$-Arg$^+$-Arg$^+$ (based on the NH_2-terminal sequence of the type II catalytic subunit of cAMP-dependent protein kinase), Gly-Ser-Ser-Lys$^+$-Ser -Lys$^+$-Pro- Lys$^+$-Arg$^+$ (based on the NH_2-terminal sequence of pp60 ^{src}), Gly-Asn-Ala-Ser-Ser-Ile-Lys$^+$- Lys-$^+$ Lys$^+$ (the NH_2-terminal sequence of the M2 gene segment of reovirus type 3), and Gly- Ala-Gln-Phe-Ser-Lys$^+$-Thr-Ala-Arg$^+$-Arg$^+$ (NH_2-terminal sequence of myristoylated alanine-rich C kinase substrate [MARCKS]. The peptide is modified from that dictated by the cAMP-dependent protein kinase sequence by the addition of two arginine residues at the C-terminus which facilitated the binding of peptide to the P81 phosphocellulose paper. The peptide Gly-Asn-Glu-Ala-Ser-Tyr-Pro-Leu (NH_2-terminal sequence of β-subunit of calcineurin) is used for alternative NMT assay method.
4. The peptides are purified by CM-cellulose column chromatography and G-25 Sephadex gel filtration ***(11)***.

3.1.2. Synthesis of Myristoyl-CoA

1. [^{3}H]labeled myristoyl-CoA is synthesized enzymatically as described by Towler and Glaser ***(6)***.
2. In the case of [1-^{14}C]myristoyl-CoA, NMT assay can be carried out directly. [1-^{14}C]myristoyl-CoA is diluted in 50 m*M* sodium acetate, pH 7.4, to obtain the desired concentrations.
3. Add 103.5 μL of myristoyl-CoA generation buffer, 20 μL of ATP, 10 μL of LiCoA, 60 μL of pseudomonas acyl-CoA synthetase, and 7.4 μL of [9,10-^{3}H]myristic acid in a total volume of 200 μL.
4. Mix the contents gently, and incubated the reaction for 30 min at 30°C in a water bath.
5. Myristoyl-CoA thus generated is used immediately.
6. The conversion of myristoyl-CoA is calculated by determining the radioactivity remaining in the aqueous phase after extracting four times with 2 vol of either heptane or petroleum ether, a modification of the method of Hosaka et al. ***(22)***. The concentration of [^{3}H]myristoyl-CoA is calculated from the percentage conversion and the specific activity of [^{3}H]myristic acid.
7. Typically, conversion of [^{3}H]myristic acid to [^{3}H]myristoyl-CoA is always >95%.
8. NMT assay is carried out in a final volume of 25 μL.
9. Add 3.85 μL of NMT assay buffer, 1.25 μL of Triton X-100, 2.5 μL of peptide, and 10 μL of crude extract as an enzyme.
10. All biological samples are kept on ice during the additions.
11. The added components are mixed, and the transferase reaction is initiated by the addition of 7.4 μL [^{3}H]myristoyl-CoA.
12. The assay tubes are incubated in water bath at 30°C for the indicated times.
13. The reaction is terminated by spotting an aliquot of the incubation mixture onto P81 phosphocellulose paper squares (2 × 2 cm), which are supported above a styrofoam board with a pin.
14. The paper squares can be dried under a heat lamp or with a hair dryer for 30 s and pins are removed.

15. The P81 phosphocellulose paper disks are transferred to washing unit **(Fig. 1)** and washed in three changes of washing buffer for 90 min (3 × 30 min).
16. The radioactivity on phosphocellulose paper disks is quantified in 7.5 mL of Beckman Ready Safe Liquid Scintillation mixture in a Beckman Liquid Scintillation Counter.
17. The assays are usually performed in duplicate, and background is always below 5% of the experimental values.
18. Under these conditions the rate of the reaction is proportional to enzyme concentration and time **(Fig. 2)**.

3.1.3. Effect of Sodium Dodecyl Sulfate on NMT Activity

SDS has been shown to activate NMT *(2)*. This activation may be owing to the opening of the structure of NMT and allows NMT to become highly active **(Fig. 3)**. If activity is very low in the desired tissue, it is recommended to use 0.05% (1.73 m*M*) SDS in the enzyme assay. However a higher concentration of SDS inactivates NMT activity **(Fig. 3)**.

3.1.4. Definition of Units and Specific Activity

One unit of enzyme activity is defined as 1 pmol of myristoylated peptide formed/min. Specific activity is expressed as U/mg of protein, and protein concentration is determined by the method of Bradford *(23)* using BSA as a standard.

3.1.5. Calculation of NMT Activity

Unit (fmol/assay):

$$\text{DPM (test} - \text{blank)} \times (25^a/15^b) \times (1/87.4) \qquad (5)$$

Specific activity of [^{3}H]myristic acid is 87.4 fmol/dpm; a = total assay volume and b = volume spotted on P81 phosphocellulose paper.

Activity (pmol/mL/min):

$$\text{U}/1000 \times 1/0.015 \times 1/\text{time} \qquad (6)$$

Specific activity (pmol/min/mg):

$$\text{Activity/protein (mg/mL)} \qquad (7)$$

3.2. Method for Assaying NMT Activity (Alternative Method)

1. A direct method is devised for assay of NMT that depends on continuously monitoring the appearance of free thiol using the thiol-specific reagent DTNB at 412 nm *(24)*.
2. Assays are performed in 96-well cluster costar ($^1/_2$ area with lid) tissue culture microtiter plates.
3. Add 10 μL of Tris-HCl buffer, pH 8.0, 5 μL peptide, 5 μL of myristoyl-CoA, and 10 μL of DTNB in a final volume of 50 μL.

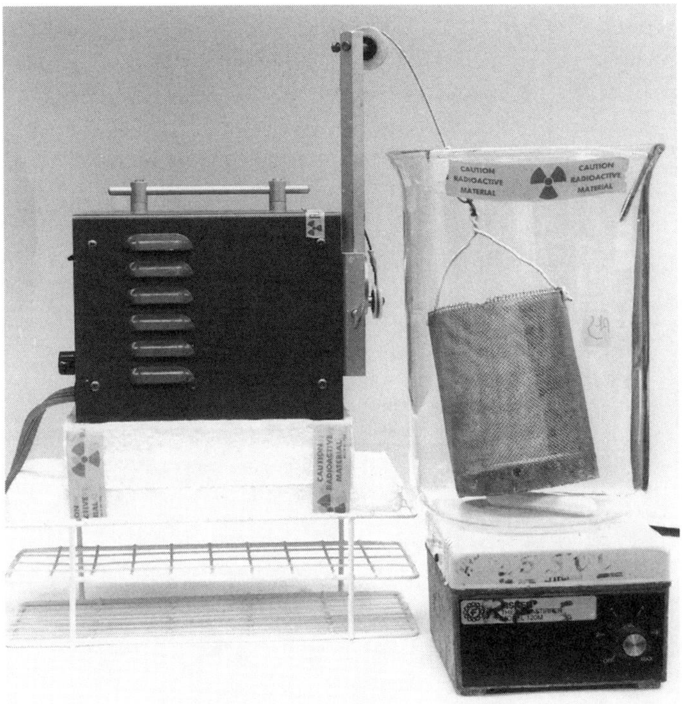

Fig. 1. Washing unit for NMT assay.

4. The transferase assay is initiated by the addition of 20 µL of NMT and incubating the reaction at 30°C.
5. The increase in absorbance at 412 nm is monitored after the addition of enzyme using either a Beckman Dual Beam Sepectrophotometer or an ELISA recorder.
6. The extension coefficient for DTNB at pH 8.0 is 13,600 M^{-1} cm^{-1} at 412 nm *(24)*, and the chromophore, 2 nitro-5-thiobenzoate is fully ionized at pH 8.0 *(25)*.
7. A standard curve relating NMT activity to CoA concentration is obtained using varying concentrations of CoA.
8. One absorbance unit at 412 nm corresponds to 0.4 nmol of CoA.
9. The specificity of this assay is examined using other saturated fatty acid coenzymes, such as palmitoyl (C16:0) and stearoyl (C18:0) CoAs **(Fig. 4)**.

3.2.1. Definition of Units and Specific Activity

One unit of enzyme activity is defined as 1 µmol of CoA released/min. Specific activity is expressed as U/mg of protein.

3.3. Method for Inhibitors of NMT Activity

1. Tissue is cut into small pieces, and is ground in a meat grinder and then homogenized in a Warring blender for 1 min in 2 vol of buffer A.

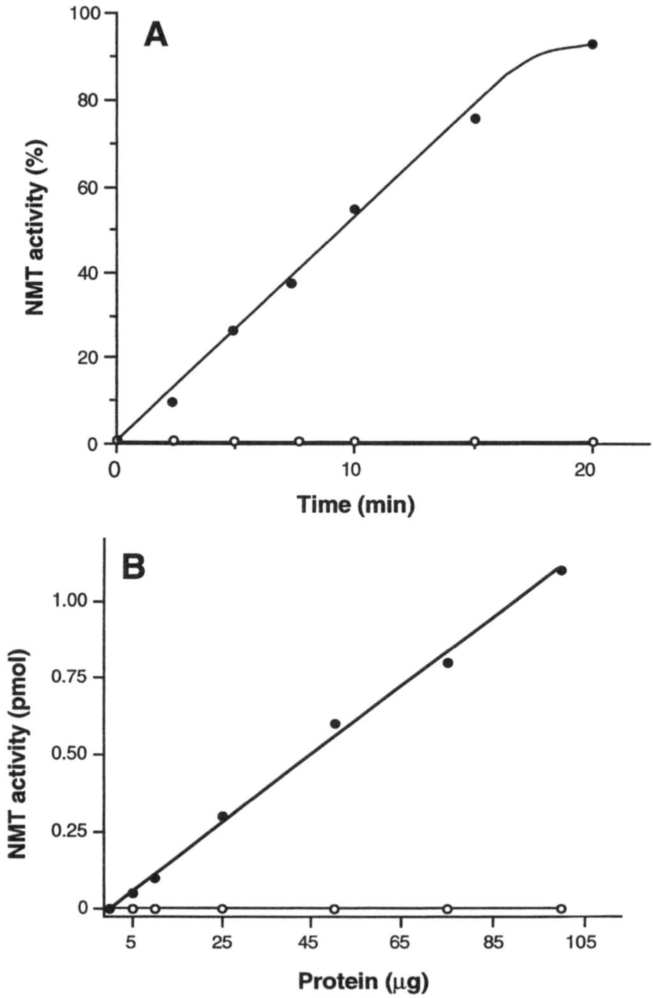

Fig. 2. Effect of time **(A)**. Bovine spleen homogenate (150–200 µg) is incubated at 30°C in either the absence (○) or presence of (●) of 20 µ*M* pp60^{src} derived peptide substrate. The reaction is initiated by the addition of 0.24 µ*M* [^{3}H]myristoyl-CoA and allowed to proceed for the indicated times. NMT activity is expressed as pmol/mg extract. Effect of protein concentration **(B)**. Bovine spleen homogenate (0–100 µg) is incubated in the absence (○) or presence of 120 µ*M* pp60^{src} derived peptide (●). The reaction is initiated by the addition of 0.24 µ*M* [^{3}H]myristoyl-CoA and allowed to proceed for 10 min at 30°C. NMT activity is expressed as pmol per assay.

2. The homogenate is centrifuged at 10,000*g* for 45 min, and the resultant supernatant is filtered through glass wool.

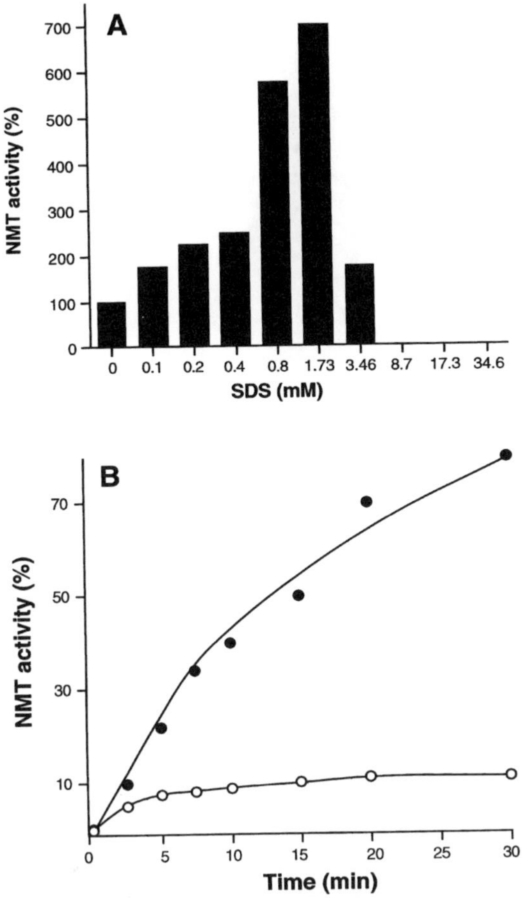

Fig. 3. Effect of SDS on bovine spleen NMT. Bovine spleen NMT activity (50 µg) is determined in the presence of varying concentrations of SDS using cAMP-dependent protein kinase-derived peptide substrate **(A)**. Time-course myristoylation of cAMP-dependent protein kinase-derived peptide by bovine spleen NMT (2 µg) in the presence (○) or absence (●) of SDS (1.73 m*M* or 0.05%) **(B)**. Results are expressed as percent of control.

3. The homogenate is centrifuged at 100,000*g* for 90 min. The subsequent supernatant is denoted the soluble (cytosolic) fraction. The pellet is resuspended in buffer A and is denoted as particulate (membrane) fraction.
4. These fractions are boiled, centrifuged, and added to the standard NMT assay for inhibitory activities.
5. Inhibitor is assayed by using the standard NMT assay as described in **Subheading 3.1.2.**, except assays are carried out with or without inhibitor at 30°C for 10 min. The standard inhibitor assay contains 0.02 pmol NMT.

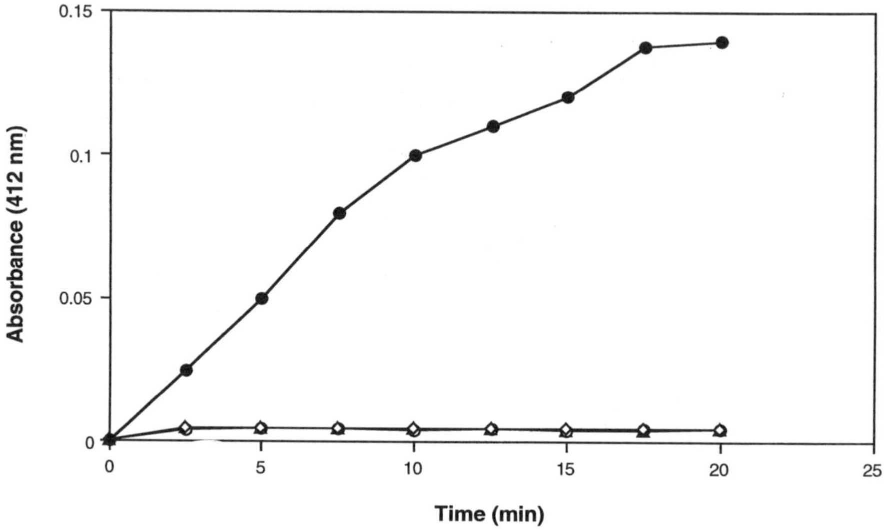

Fig. 4. Fatty acid specificity of NMT. Bovine spleen NMT (50 μg) is incubated in the presence of 25 m*M* Tris-HCl buffer, pH 8.0, 500 μ*M* pp60[src], 1 m*M* DTNB, and 100 μ*M* myristoyl-CoA (●), palmitoyl-CoA (○), stearoyl-CoA (▲), and NMT in the absence of peptide (△) at 26°C . The release of CoA is monitored at 412 nm.

3.3.1. Definition of Units and Specific Activity

One unit of NMT inhibitor activity is defined as U/mg, where 1 U of inhibitor protein inhibits the formation of 1 pmol of myristoylated peptide/min.

3.4. Method for Analysis of NMT-Protein Expression

3.4.1. Synthesis of NMT–Peptide Conjugates

1. The peptide NENYVEDDDNMFRFD corresponding to the amino-terminal region from amino acids 97–112 of human NMT *(26)* is synthesized (Applied Biosystems Model 430 A peptide synthesizer, Alberta Peptide Institute, Canada) utilizing *t*-Boc *N*α-protection and benzyl-type side-chain protection.
2. The peptide is coupled to two carrier proteins, BSA (peptide-to-protein ratio 14:1) and KLH, peptide-to-protein ratio 10:1) *(27)*.

3.4.2. Solubilization of KLH–Peptide Conjugate

1. Dissolve 2 mg of KLH–peptide conjugate in 1 mL of 8 *M* urea in a 1.5-mL Eppendorf tube with occasional vortexing over a period of 4 h.
2. Centrifuged for 5 min, and decant the supernatant into a 6800-mol-wt cutoff dialysis bag.
3. If a pellet is observed, 50 μL of glacial acetic acid are added with further occasional vortexing over a period of 30 min.

4. Five hundred microliters of 8 *M* urea are added to the mixture and vortexed.
5. The KLH conjugate is dialyzed against 0.9% NaCl overnight.

3.4.3. Antipeptide Antibody Production

1. Antigen (500 µg) KLH conjugate is mixed with 0.5 mL Freund's complete adjuvant and injected into rabbits on day 0 (0.25 mL under each shoulder blade and 0.5 mL in one hip muscle [subscapular and gluteal injections]).
2. Subsequently on days 21, 35, and 49, 500 µg of antigen are mixed with Freund's incomplete adjuvant and injected.
3. Blood is collected 10 d after each injection from a marginal ear.
4. The BSA–peptide conjugate is used to screen for antipeptide antibody by EIA ***(28)***.

3.4.4. Preparation of BSA–Peptide Affinity Column

1. Weigh out the required amount of freeze-dried powder of activated CH Sepharose-4B (1 g activated CH Sepharose-4B gives approx 3 mL swollen gel) suspended in 1 m*M* HCl.
2. The gel swells immediately. Wash for 15 min with 1 m*M* HCl on a sintered glass filter. Use approx 200 mL/g freeze dried powder.
3. Dissolve 10 mg of BSA–peptide in 10 mL of 0.1 *M* $NaHCO_3$, pH 9.6, and 2 g of activated CH Sepharose-4B.
4. The resultant slurry is mixed by rotation for 16 h at 4°C.
5. The Sepharose-4B is pelleted by centrifugation (1000*g* for 5 min), and the extent of BSA–peptide coupling is determined by measuring the protein concentration of the peptide conjugate in the supernatant.
6. The BSA–peptide–Sepharose-4B is washed with 0.1 *M* $NaHCO_3$, pH 9.6, in a Buchner flask.
7. Any remaining reactive groups are blocked by incubating the gel with 0.1 *M* Tris-HCl, pH 7.4, for 2 h at room temperature.
8. The BSA–peptide–Sepharose-4B is washed alternately with low-pH (0.1 *M* acetate buffer, pH 4.0) and high-pH (0.1 M Tris-HCl, pH 8.0) buffers containing 0.5 *M* NaCl. Gel is ready for use.

3.4.5. Affinity Purification of Antipeptide Antibodies

1. Apply antisera to a 5-mL BSA–peptide–Sepharose-4B (10 × 1.5 cm) column.
2. After adsorption, the column is washed with 0.1 *M* Tris-HCl, pH 8.0, containing 0.5 *M* NaCl.
3. The antipeptide antibodies are eluted using 100 m*M* glycine pH 2.5.
4. One-milliliter fractions are collected in tubes containing 3.5 mL of 1 *M* Tris-HCl at pH 8.0 and mixed immediately.
5. The purified antipeptide antibodies are concentrated by dialysis against 30% polyethylene glycol, suspended in 20 m*M* Tris-HCl, pH 7.0, 50% glycerol, and 0.02% NaN_3, and then stored at –70°C at a protein concentration of 0.9 mg/mL in small aliquots.

3.4.6. Enzyme Immunoassay (EIA)

1. EIA is performed in 96-well, flat-bottom plates (Becton Dickinson Company, Lincoln Park, NJ) in triplicate wells.
2. Each well is coated with 100 µL of NMT in carbonate-bicarbonate buffer (0.05 *M*, pH 9.6), and plates are left at 4°C for overnight, and washed three times with 0.01 *M* PBS, pH 7.4, containing 0.5% BSA Fraction V (Sigma).
3. The wells are blocked with 250 µL of PBS containing 5% skimmed milk powder for 2 h at 37°C.
4. Following three more washes, 100 µL of the primary antibody (antipeptide antibody [1:12,500]) is applied to each well and incubated for 2 h at 37°C.
5. After five washes, 100 µL of peroxidase-conjugated goat antirabbit IgG (Bio-Rad Laboratories Inc., Hercules, CA) at a dilution of 1:2000 in PBS is added and incubated for another 2 h at 37°C.
6. Finally, the plates are washed five times with PBS and 125 µL of horseradish peroxidase conjugate substrate solution containing premixed 4-chloro-1-naphthol and hydrogen peroxide (Bio-Rad Laboratories Inc.) is added.
7. The plates are then incubated at room temperature in the dark.
8. After 15 min, the reaction is stopped by adding 125 µL of 4 m*M* aqueous sodium azide, and the optical density is read at 405 nm.
9. Antibody specificity was carried out using 10 times excess of peptide conjugate.
10. NMT levels in the samples are determined from a standard curve obtained by plating known concentrations of purified human NMT *(29)*.
11. One absorbance of the reaction at 415 nm corresponds to 8.5 ng of purified human NMT protein.
12. Background is evaluated by using preimmune serum.

3.5. Method for Purification of NMT from Bovine Spleen

3.5.1. Extraction

1. All procedures are performed at 4°C unless otherwise stated.
2. Fresh bovine spleens are obtained from the local slaughterhouse and transferred to the laboratory in packed ice.
3. Spleen tissue (500 g) is cut into small pieces and is ground in a meat grinder, and then homogenized in a Warring blender for 1 min in 2 vol of buffer A.
4. The homogenate is centrifuged at 10,000*g* for 45 min, and the resultant supernatant is filtered through glass wool.

3.5.2. Ammonium Sulfate Fractionation

1. Solid ammonium sulfate (21g/100 mL) is added slowly to the supernatant to achieve 35% saturation.
2. The supernatant is gently stirred for 25–30 min and then centrifuged at 10,000*g* for 30 min. The pellet is discarded.
3. To the supernatant additional solid ammonium sulfate (24 g/100 mL) is added to achieve 70% saturation and then centrifuged at 10,000*g* for 30 min.

4. The resultant protein pellet is dissolved in buffer A. This enzyme preparation is stable for several months when stored at –70°C.

3.5.3. SP-Sepharose Fast-Flow Column Chromatography

1. Ammonium sulfate fraction (35–70%) is applied to an SP-Sepharose fast-flow column (8.5 × 20 cm) which is pre-equilibrated with buffer B.
2. After application, the column is washed with 2 bed volumes of buffer B.
3. The column is further washed with 1.5 column volumes of buffer B containing 75 m*M* NaCl until no protein is detected by the dye binding method ***(14)***.
4. The bound protein is eluted with buffer B containing 1 *M* NaCl and 10% (w/v) sucrose.
5. Fractions (20 mL) are collected. Fractions containing NMT activity are pooled.

3.5.4. Phenyl Sepharose CL-4B Column Chromatography

1. The pooled SP-Sepharose sample is applied to a phenyl-Sepharose CL-4B column (2.5 × 9.0 cm), which was pre-equilibrated previously with buffer C.
2. After application, the column is washed with 10 column volumes of buffer C until no protein is detected.
3. Most of the protein in the original sample passes through without binding.
4. No NMT activity is detected in the initial elution with buffer C containing 1 *M* NaCl.
5. NMT activity is eluted with buffer C containing 80% (v/v) ethylene glycol.

3.5.5. SP-Sepharose Fast-Flow Column Chromatography

1. A SP-Sepharose column is used to concentrate NMT into a small volume and also to remove ethylene glycol.
2. Pooled sample from the last step is applied to an SP-Sepharose column (2.5 × 10.5 cm) pre-equilibrated with buffer B.
3. The column is washed with 5 bed volumes of buffer B containing 75 m*M* NaCl.
4. The NMT activity is eluted with buffer B containing 1 *M* NaCl and 10% (w/v) sucrose.
5. The NMT activity is dialyzed overnight against buffer D.

3.5.6. DEAE-Sepharose CL-6B Column Chromatography

1. Dialyzed NMT activity is applied to a DEAE-Sepharose CL-6B column (1.5 × 13 cm), which is pre-equilibrated with buffer D.
2. The NMT activity failed to bind to the column or was eluted with buffer D containing 50 m*M* NaCl.
3. The unbound fraction and initial column wash from DEAE-Sepharose CL-6B are pooled and concentrated by ultrafiltration through an Amicon YM 10 membrane.
4. The majority of the bound protein, eluted with buffer D containing 1 *M* NaCl, does not possess NMT activity.

3.5.7. Superose 12 (HR/30) FPLC Gel Filtration

1. Five hundred microliters of the unbound concentrated NMT fraction from DEAE-Sepharose CL-6B column are applied to a Superose gel-filtration FPLC pre-equilibrated with buffer B containing 0.15 *M* NaCl and 5% (w/v) sucrose.
2. Fractions are collected at a flow rate of 0.5 mL/min. The NMT activity appears as a sharp peak.
3. Fractions having high NMT activity are pooled, and SDS-polyacrylamide gel electrophoresis is carried out to examine the purity (**Fig. 5**, inset).
4. The pooled samples are stored in small aliquot at –70°C until use.

3.6. Method for Demyristoylation of Peptide Substrates

1. NMT assays are carried out in a manner similar to that presented in **Subheading 3.1.2.** except for the generation of myristoyl-CoA with low (1.12 μ*M*) and high (35 μ*M*) concentrations of LiCoA.
2. Myristoyl-CoA generated with 1.12 μ*M* CoA involved myristoylation during early phase of reaction (**Fig. 6**), however, demyristoylation occurred at later stage of the reaction.
3. In the presence of 35 μ*M* CoA, the reaction involved exclusively myristoyl-peptide synthesis (**Fig. 6**).

4. Notes

1. NMT assay is simple and more convenient for handling multiple samples.
2. [^{3}H]myristoyl-peptide was separated from unreacted [^{3}H]myristic acid and [^{3}H]myristoyl-CoA by washing the P81 phosphocellulose paper disks at pH 7.3.
3. The expected charges on the myristoylated peptides at this pH are +4 and +3 for the peptides, which facilitates their binding to the negatively charged P81 phosphocellulose paper matrix. Both the myristic acid and myristoyl-CoA would be negatively charged at this pH. Consequently, these reactants do not bind to the phosphocellulose paper.
4. The only limitation with this assay is that the peptides should have positive charges at the C-terminal end, which facilitate the binding of peptides to phosphocellulose paper.
5. The spectrophotometric assay is simple and convenient for carrying out multiple samples.
6. This method does not involve the use of radiolabeled compounds and positive charged peptides.
7. Recently, we have reported that NMT activity is high in animal and human colorectal adenocarcinomas ***(30)***.
8. EIA may be an alternative method to measure NMT expression in tumors ***(31)***/ tissues when small amounts of tissue/biopsy samples are available.
9. Another distinct advantage of EIA is that when combined with the NMT activity assay, it gives an indication of any NMT activators or inhibitors in tissues. For

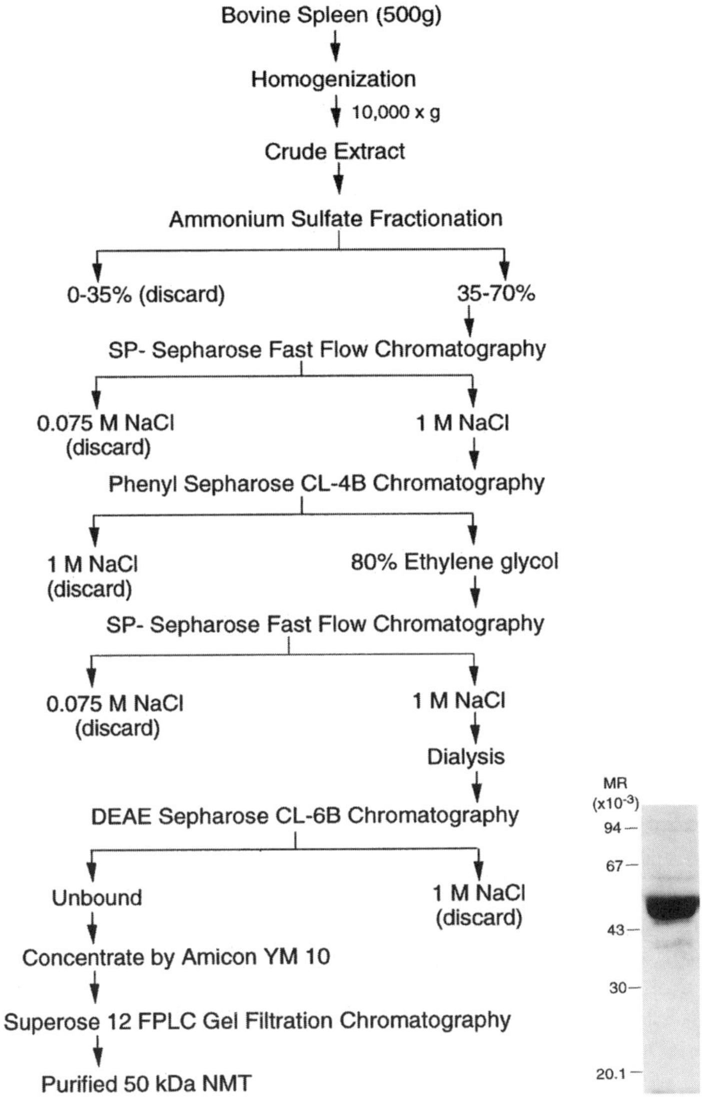

Fig. 5. Flowchart of bovine spleen NMT purification. Inset: SDS-PAGE is carried out with 20 µg of bovine spleen NMT. Molecular-weight markers used in this study are phosphorylase b (94 kDa), BSA (67 kDa), ovalbumin (43 kDa), carbonic anhydrase (30 kDa), and trypsin inhibitor (20.1 kDa).

example, if the expression is not correlating with the activity, this suggests the presence of endogenous NMT modulators in the tissues.

10. Antibody is useful to visualize NMT in frozen tissue sections.

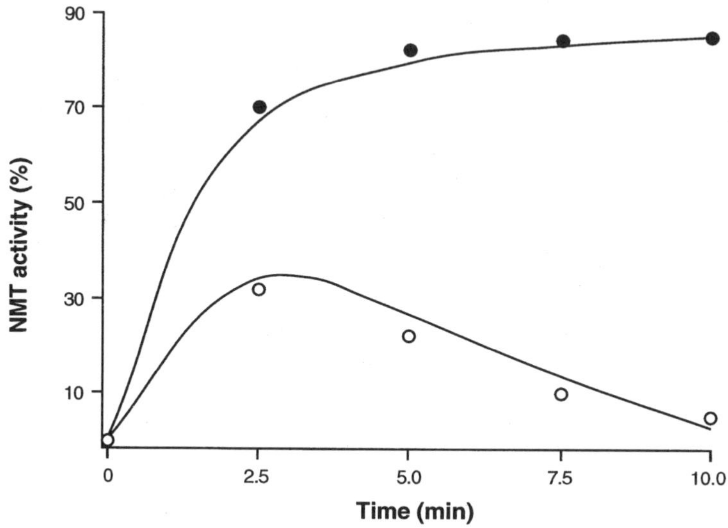

Fig. 6. Effect of various concentrations of CoA on bovine spleen NMT. Bovine spleen NMT (0.23 μg) is incubated with cAMP-dependent protein kinase-derived peptide substrate (500 μ*M*) using myristoyl-CoA generated from 1.12 μ*M* (○) and 35 μ*M* (●).

11. Bovine brain NMT has been reported to be upregulated by a cytosolic activator and downregulated by a membrane-bound inhibitor.
12. Identification of these inhibitors in various tissues/cell types would give a better understanding of the role of these inhibitors in NMT regulation/protein myristoylation.
13. The purification procedure for NMT from bovine spleen is described as a flowchart in **Fig. 5**.
14. The purification procedure is reproducible, rapid, and suitable for large-scale preparation. NMT activity is usually purified between 1475- and 1800-fold from crude spleen extract with 23–30% yield **(Table 1)**.
15. The specific activity of the purified NMT sample is in the range of 41,000–50,000 U/mg. The purified enzyme preparation is stable for at least 12 mo when stored at –70°C.
16. The purified bovine spleen NMT has an apparent molecular mass of 58 kDa as determined on a calibrated gel-filtration FPLC column ***(15)***. SDS-PAGE of the NMT showed a single protein band with an apparent molecular mass of 50 kDa (**Fig. 5**, inset). These results suggest that the NMT is a monomeric enzyme.
17. Based on our results, we hypothesized that the initial event in the regulation of NMT can be an increase in CoA concentration in the cell, which is coupled to a rise in protein myristoylation. Once the CoA concentration decreases in the cell, the demyristoylation reaction becomes operative. These results suggested that the CoA concentration may modulate the myristoylation reaction in vivo.

Table 1
Purification of NMT from Bovine Spleen[a]

Fraction	Total protein, mg	Specific activity, U/mg	Purification, -fold	Yield, %
35–70% $(NH_4)_2SO_4$	5152	28	1	100
SP-Sepharose fast flow column	192	219	8	29
Phenyl-Sepharose CL-4B column	21	2933	105	43
DEAE-Sepharose CL-6B column (unbound)	3.6	12,204	439	30
Superose-12 FPLC column	0.8	41,000	1475	23

[a]One unit of NMT activity is defined as 1 pmol of myristoyl peptide formed/min.

Acknowledgments

This work is supported by the Medical Research Council of Canada and the Heart and Stroke Foundation of Saskatchewan, Canada. R. V. S. Raju is a recipient of a Research Fellowship from the Health Services Utilization and Research Commission of Saskatchewan.

References

1. Boutin, J. (1997) Myristoylation. *Cell Signal.* **9,** 15–35.
2. Raju, R. V. S., Magnuson, B. A. and Sharma, R. K. (1995) Mammalian myristoyl-CoA:protein *N*-myristoyltransferase. *Mol. Cell. Biochem.* **149/150,** 191–202.
3. Carr, S. A., Biemann, K., Shoji, S., Parmelee, D. C., and Titani, K. (1982) *n*-tetradeconyl is the NH_2-terminal blocking group of the catalytic subunit of cyclic AMP-dependent protein kinase from bovine cardiac muscle. *Proc. Natl. Acad. Sci. USA* **79,** 6128–6131.
4. Aitken, A., Cohen, P., Santikaran, S., Williams, D. H., Calder, A. G., Smith, G. et al. (1982) Identification of NH_2-terminal blocking group of calcineurin B as myristic acid. *FEBS. Lett.* **150,** 314–318.
5. Johnson, D. R., Bhatnagar, R. S., Knoll, L. J., and Gordon, J. I. (1994) Genetic and biochemical studies of protein *N*-myristoylation. *Ann. Rev. Biochem.* **63,** 869–914.
6. Towler, D. A. and Glaser, L. (1986) Protein fatty acid acylation. Enzymatic synthesis of an *N*- myristoylglycyl peptide. *Proc. Natl. Acad. Sci. USA* **83,** 2812–2816.
7. Towler, D. A., Adams, S. P., Eubanks, S. H., Towery, D. S., Adams, S. P., and Glaser, L. (1987) Amino terminal processing of proteins by *N*-myristoylation: Substrate specificity of *N*-myristoyltransferase. *J. Biol. Chem.* **262,** 1030–1036.
8. Paige, L. A., Chafin, D. R., Cassady, J. M., and Geahlen, R. L. (1989) Detection of myristoyl- CoA:protein *N*-myristoyltransferase activity by ion-exchange chromatography. *Anal. Biochem.* **181,** 254–258.
9. McIlhinney, R. A. J. and McGlone, K. (1989) A simple assay for the enzyme responsible for the attachment of myristic acid to the *N*-terminal glycine residue

of proteins, myristoyl-CoA:glycylpeptide *N*-myristoyltransferase. *Biochem. J.* **263,** 387–391.
10. French, S. A., Christakis, H., O'Neil, R. R., and Miller, S. P. (1994) An assay for myristoyl- CoA:protein *N*-myristoyltransferase activity based on a ion-exchange exclusion of [^{3}H]myristoyl peptides. *Anal.Biochem.* **220,** 115–121.
11. King, M. J. and Sharma, R. K. (1991) *N*-myristoyltransferase assay using phosphocellulose paper binding. *Anal. Biochem.* **199,** 149–153.
12. Rudnick, D. A., McWherter, C. A., Adams, S. P., Ropson, I. J., Duronio, R. J., and Gordon, J. I. (1990) Structural and functional studies of *Saccharomyces cerevisiae* myristoyl-CoA:protein *N*-myristoyltransferase produced in *Escherichia coli*. Evidence for an acyl-enzyme intermediate. *J. Biol. Chem.* **265,**13,370–13,378.
13. Towler, D. A., Adams, S. P., Eubanks, S. R., Towery, D. S., Jackson-Machelski, E., Glaser, L., et al. (1988) Myristoyl-CoA:protein *N*-myristoyltransferase activities from rat liver and yeast possess overlapping yet distinct peptide substrate specificities . *J. Biol. Chem.* **263,** 1784–1790.
14. McIlhinney, R. A. J. and McGlone, K. (1990) Characterization of myristoyl-CoA:glycylpeptide *N*-myristoyltransferase activity in rat brain. Subcellular and regional distribution. *J. Neurochem.* **54,** 110–117.
15. Raju, R. V. S., Kalra, J., and Sharma, R. K. (1994) Purification and properties of bovine spleen myristoyl-CoA:protein *N*-myristoyltransferase. *J. Biol. Chem.* **269,** 12,080–12,083.
16. Raju, R. V. S. and Sharma, R. K. (1995) Myristoyl-CoA:protein *N*-myristoyltransferase. Subcellular localization, activation and kinetic behaviour in the presence of organic solvents. *Biochem. Biophys. Res. Commun.* **208,** 617–623.
17. Raju, R. V. S., Datla, R. S. S., and Sharma, R. K. (1996) Reduction of oncoprotein transformation in vitro by albumin. *J. Natl. Cancer. Inst.* **88,** 556–557.
18. King, M. J. and Sharma, R. K. (1993) Identification, purification and characterization of membrane-associated *N*-myristoyltransferase inhibitor protein from bovine brain. *Biochem. J.* **291,** 635–639.
19. Swarup, G., Dasgupta, J. D., and Garbers, D. (1983) Tyrosine protein kinase activity of rat spleen and other tissues. *J. Biol. Chem.* **258,** 10,341–10,347.
20. Raju, R. V. S. and Sharma, R. K. (1996) Coenzyme A dependent myristoylation and demyristoylation in the regulation of bovine spleen *N*-myristoyltransferase. *Mol. Cell. Biochem.* **158,** 107–113.
21. Merrifield, R. B. (1986) Solid phase peptide synthesis. *Science* **232,** 341–347.
22. Hosaka, K., Mishina, M., Kamiryo, T., and Numa, S. (1981) Long-chain acyl-CoA synthetases I and II from *Candia lipolytica*. *Methods. Enzymol.* **71,** 325–333.
23. Bradford, M. M. (1976) A rapid and sensitive method for the quantitation of microgram quantities of protein utilizing the principle protein-dye binding. *Anal. Biochem.* **72,** 248–254.
24. Ellman, G. L. (1959) Tissue sulfhydryl groups. *Arch. Biochem. Biophys.* **82,** 70–77.
25. Riddles, P. W., Blakeley, R. L., and Zerner, B. (1983) Reassessment of Ellman's reagent. *Methods. Enzymol.* **91,** 49–60.

26. Duronio, R. J., Reed, S. I., and Gordon, J. I. (1992) Mutations of human myristoyl-CoA:protein *N*-myristoyltransferase cause temperature sensitive myristic acid auxotrophy in *Saccharomyces cerevisiae*. *Proc. Natl. Acad. Sci. USA* **89,** 4129–4133.
27. Harlow, E., and Lane, D. (eds.) (1988) Antibodies. A Laboratory Manual. Cold Spring, Harbor Laboratory Press, Cold Spring Harbor, NY.
28. Voller, A., Bartlett, A. and Bidwell, D. E. (1978) Enzyme immuno assay with special reference to ELISA techniques. *J. Clin. Pathol. (Lond.)* **31,** 507–520.
29. Raju, R. V. S., Datla, R. S. S., and Sharma, R. K. (1996) Overexpression of human *N*- myristoyltransferase utilizing a T7 polymerase gene expression system. *Protein Expression Purif.* **7,** 431–437.
30. Magnuson, B. A., Raju, R. V. S., Moyana, T. N., and Sharma, R. K. (1995) Increased *N*- myristoyltransferase activity observed in rat and human colonic tumors. *J. Natl. Cancer. Inst.* **87,** 1630–1635.
31. Raju, R. V. S., Moyana, T. N., and Sharma, R. K. (1997) *N*-myristoyltransferase overexpression in human colorectal adenocarcinomas. *Exp. Cell. Res.* **235,** 145–154.

16

Metabolic Labeling of Protein-Derived Lipid Thioesters in Palmitoyl-Protein Thioesterase-Deficient Cells

Sandra L. Hofmann and Jui-Yun Lu

1. Introduction

Palmitoyl-protein thioesterase (PPT) is a lysosomal enzyme that removes fatty acids bound in thioester linkage to cysteine residues in fatty acylated proteins *(1–4)*. Mutations in the PPT gene that affect enzyme activity cause a recessively inherited lysosomal storage disorder, infantile neuronal ceroid lipofuscinosis (INCL) *(5)*. The disorder is characterized by severe central nervous system degeneration and lysosomal inclusions in the form of granular osmiophilic deposits in many tissues *(6,7)*.

Cells derived from INCL subjects accumulate lipophilic thioesters derived from the lysosomal degradation of fatty acylated proteins *(8,9)*. To demonstrate these abnormally accumulating thioesters in INCL cells, newly synthesized cellular proteins are labeled with [^{35}S]cysteine. With time, a small proportion of these protein-incorporated cysteine residues will become acylated and eventually travel with the protein to lysosomes for degradation. In PPT-deficient cells, the acyl-cysteine bond is not cleaved, and the long-chain fatty acid imparts a hydrophobic character to a proportion of the [^{35}S]cysteine-labeled material. This material is extracted from the labeled cells using a simplified chloroform/methanol extraction based on the Bligh-Dyer procedure. This method has been shown to extract quantitatively acyl CoAs from cultured cells *(10)*. The radiolabeled compounds in the organic phase are analyzed by thin-layer chromatography (TLC) as shown in **Fig. 1**. They can be further subjected to chemical hydrolysis using neutral hydroxylamine or enzymatically hydrolyzed using purified recombinant PPT.

From: *Methods in Molecular Biology, Vol. 116: Protein Lipidation Protocols*
Edited by: M. H. Gelb © Humana Press Inc., Totowa, NJ

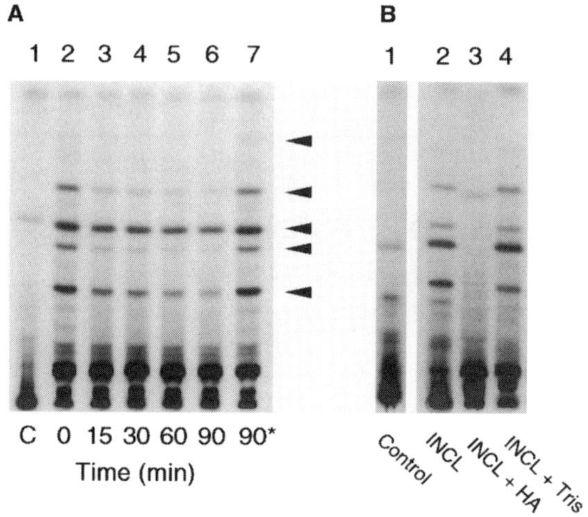

Fig. 1. Hydrolysis of [^{35}S]cysteine-labeled lipids in vitro by PPT **(A)** and neutral hydroxylamine **(B)**, and analysis by HP-TLC and fluorography. Lipid extracts cor-responding to 5×10^5 INCL lymphoblasts/lane (5000 cpm) were incubated with 0.25 µg of recombinant bovine PPT at 37°C for the times indicated in **(A),** lanes 2–6. **(A)** Lanes: 1, lipid extract of normal control (C) lymphoblasts; 7, lipids incubated with heat-denatured PPT. **(B)** Neutral hydroxylamine treatment of [^{35}S]cysteine-labeled lipids from INCL lymphoblasts. Lanes: 1, normal control lipid extract, untreated; 2, INCL lipid extract, untreated; 3, INCL lipid extract, incubated for 1 h at 23°C with 1 *M* hydroxylamine, pH 8.0; 4, INCL lipid extract incubated with 1 *M* Tris-HCl, pH 8.0. Reprinted from ref. *(8)* with permission (Copyright [1996] Natl. Acad. of Sci. USA.)

We have successfully applied these methods to analyze the abnormal cysteine-containing lipid thioesters in both transformed lymphoblasts and fibroblasts in culture. Below is described the basic method as applied to lymphoblasts derived from INCL subjects, which normally grow in suspension culture. Deviations from the method as applied to fibroblasts are noted.

2. Materials

2.1. Cell Labeling with [^{35}S]cysteine

1. RPMI 1640 (Gibco, Gaithersburg, MD, cat. no. 11875-01) tissue-culture medium supplemented with 10% heat-inactivated fetal bovine serum, penicillin (100 U/mL), streptomycin (100 µg/mL), and amphotericin B (0.25 µg/mL) (Gibco, cat. no. 15240-062).
2. Cysteine- and methionine-free RPMI 1640 tissue-culture medium (Cellgro Mediatech, Herndon, VA, cat. no. 17-104-LI), to which supplemental methionine (Sigma, St. Louis, MO, cat. no. M2893) is added (final concentration, 15 µg/mL).

3. [^{35}S]Cysteine (1231 Ci/mmol) (ICN, Costa Mesa, CA).
4. Incubator suitable for tissue culture set at 37°C.
5. Polypropylene tubes (50 mL).
6. Bench centrifuge.
7. 1.5-mL microfuge tubes.
8. Microcentrifuge (Eppendorf centrifuge 5402) or equivalent.
9. Phosphate-buffered saline (PBS), Dulbecco's, without calcium or magnesium chloride (Gibco, cat. no. 14190-144).
10. Cell scraper.

2.2. Extraction

1. PBS, Dulbecco's, without calcium or magnesium chloride (Gibco, cat. no. 14190-144).
2. Chloroform/methanol (1:1, v/v).
3. 1.5-mL microfuge tubes.
4. Vortex mixer.
5. Microcentrifuge (Eppendorf centrifuge 5402) or equivalent.
6. Benzene.
7. Oxygen-free nitrogen gas (zero-grade).
8. N-EVAP evaporator, nitrogen-delivery type (Organomation Associates, Berlin, MA).
9. Scintillation counter.
10. Scintillation vials.
11. Scintillation fluid (3a70B, Research Products International, Prospect, IL).

2.3. High Performance Thin-Layer Chromatography (HP-TLC)

1. Solvent system: chloroform:methanol:water (65:25:4 [v/v/v]).
2. Silica gel 60 HP-TLC plates, glass-backed, 10 × 10 cm (EM Science, Gibbstown, NJ).
3. Pencil.
4. Ruler.
5. Chromatography tank with filter paper lined on the narrow sides and air-tight lid.
6. Vacuum grease.
7. Glass capillary tubes.
8. EN^{3}HANCE spray (Dupont, Wilmongton, DE).
9. Film (Kodak, Rochester, NY, X-omat)

2.4. Chemical Hydrolysis of Lipid Thioesters

1. Solvent resistant microfuge tubes (1.5 mL) (polypropylene, Robbins Scientific, Sunnyvale, CA, or equivalent).
2. Nitrogen gas (zero-grade).
3. Vortex mixer.
4. 1 *M* hydroxylamine, adjusted to pH 8.0 with KOH.
5. 1 *M* Tris-HCl, pH 8.0.

2.5. Enzymatic Hydrolysis of Lipid Thioesters

1. Solvent resistant microfuge tubes (1.5 mL) (polypropylene, Robbins Scientific or equivalent).

2. Nitrogen gas (zero-grade).
3. Vortex mixer.
4. Heated bead bath set at 100°C.
5. Purified recombinant bovine PPT *(2)*.
6. Water bath set at 37°C.
7. Nitrogen gas (zero-grade).
8. Reaction buffer: 50 m*M* Tris-HCl, pH 7.0, 50 m*M* NaCl, 1 m*M* EDTA.

3. Methods

3.1. Metabolic Labeling

1. Centrifuge 25 mL of logarithmic-phase lymphoblasts in suspension culture (approximately 7.5×10^5 cells/mL, *see* **Note 1**) in a 50-mL polypropylene tube for 5 min at $220 \times g_{max}$. For metabolic labeling of fibroblasts, *see* **Note 2**.
2. Add 5 mL of prewarmed cysteine- and serum-free medium, resuspend the cells (*see* **Note 3**), and centrifuge again at 220 g_{max}. Repeat once.
3. Resuspend the entire cell pellet in 0.5 mL of cysteine-free RPMI medium, and transfer to a 35-mm culture dish. Incubate the cells for 30 min at 37°C to deplete intracellular cysteine stores.
4. Add 0.5 mL of medium containing [^{35}S]cysteine (final concentration, 50 µCi/mL). Incubate for 4–6 h for maximal incorporation of counts.
5. Gently dislodge the cells with a rubber policeman, and pellet briefly at 14,000 rpm in a microcentrifuge tube at 4°C. (Start centrifuge and stop run as soon as it comes to speed. Total time, approx 10 s.)
6. Add 1 mL of ice-cold PBS, resuspend the cells, and pellet again for 10 s at 14,000 rpm in a microcentrifuge at 4°C.

3.2. Extraction

1. Resuspend the labeled cell pellet in 100 µL of ice-cold PBS in a microfuge tube.
2. Add 6 vol (600 µL) of chloroform/methanol, 1:1 (v/v). Vortex vigorously for 20 s and centrifuge in microcentrifuge for 5 min at 2000 rpm (400 g_{max}).
3. Carefully remove the organic (lower) phase, and place in a clean tube. Add 1/6 vol of PBS, vortex, and centrifuge for 5 min at 2000 rpm.
4. Withdraw the organic (lower) phase to a clean tube. Add an equal volume of benzene, and dry under a stream of nitrogen.
5. Add 100 µL of 1:1 chloroform/methanol (v/v). Withdraw 5 µL, and count in a liquid scintillation counter (to achieve more accurate pipet volumes, first equilibrate pipet with solvent by pipeting up and down in solvent before withdrawing desired volume). Adjust the volume to achieve a concentration of 500 cpm/µL (typical yield, 50,000–100,000 cpm). Store the sample under N_2 in a glass vial at –20°C.

3.3. High-Performance Thin-Layer Chromatography (HP-TLC)

1. Add chloroform:methanol:water (65:25:4 [v/v/v]) to a glass chromatography tank (which has been lined with absorbent paper on the narrow sides) to a

depth of 0.5 cm (about 100 mL). Rim the lid with grease to seal, and equilibrate for 2 h.

2. Mark an HP-TLC plate with pencil by drawing across the plate 1.2 cm from the bottom (sample application) and 1 cm away from the top (solvent front).
3. Apply sample (20 µL, or 10,000 cpm/lane) using a fine glass capillary. (For samples derived from fibroblasts, *see* **Note 4**). Make even streaks parallel to the solvent front and 0.4 cm in length. Avoid scratching the silica with the glass tip. Allow the applied sample to dry.
4. Rinse the capillary tube with chloroform/methanol several times, and repeat **step 3** for the other samples leaving 0.5 cm between samples. Air-dry the plate for at least 10 min.
5. Place the plate in the solvent with the uncoated side facing toward the glass wall of the tank.
6. When the solvent has reached to within 1 cm of the top of the plate (about 25 min), remove and allow to air-dry in a fume hood (*see* **Note 5**).
7. When the plate is free of all traces of solvent, spray with EN^3HANCE, and allow to dry in the fume hood for 20 min.
8. Apply the plate directly to X-ray film and expose at –85°C. A sample loaded at 5000 cpm/lane will require at least 2 d for optimal results. Results may be quantified by use of a phosphorimager, if available.

3.4. Chemical Hydrolysis of Lipid Thioesters Using Neutral Hydroxylamine

1. Add lipid extract containing approximately 40,000 cpm (in $[^{35}S]$) to each of two microfuge tubes, and evaporate the solvent under a stream of nitrogen.
2. Resuspend in 50 µL of 1 *M* hydroxylamine, pH 8.0, or 1 *M* Tris-HCl, pH 8.0 (as a control) with vigorous mixing.
3. Incubate at ambient temperature for 1 h.
4. Add 300 µL of chloroform/methanol 1:1 (v/v), and vortex vigorously.
5. Centrifuge at 2000 cpm (400 g_{max}) for 5 min to separate phases.
6. Remove the lower (organic) phase to a clean microfuge tube.
7. Add 1 vol of benzene to the organic phase. Evaporate to dryness under nitrogen.
8. Redissolve the sample in 60 µL of chloroform/methanol, 1:1 (v/v).
9. Analyze 15 µL (about 5000 cpm) by HP-TLC as described under **Subheading 3.3.**

3.5. Enzymatic Hydrolysis of Lipid Thioesters Using Recombinant PPT

1. Add lipid containing 20,000 cpm (in $[^{35}S]$) to each of two microfuge tubes, and evaporate the solvent under a stream of nitrogen.
2. Add 50 µL of reaction buffer containing 50 m*M* Tris-HCl, pH 7.0, 50 m*M* NaCl, and 1 m*M* EDTA to each tube, and vortex vigorously.
3. Heat-inactivate 1 µg of recombinant PPT at 100°C for 5 min.
4. Add 1 µg of PPT or heat-inactivated PPT.

5. Incubate samples at 37°C for 1 h. Stop reaction by placing tubes on ice and add 600 µL of chloroform/methanol/benzene 1:1:2 (v/v).
6. Evaporate to dryness under nitrogen, and redissolve in 60 µL of chloroform/methanol, 1:1 (v/v).
7. Analyze 15 µL (5,000 cpm) by HP-TLC as described under **Subheading 3.3.** above.

4. Notes

1. Immortalized lymphoblastoid cell lines are routinely maintained in a 5% CO_2 incubator at 37°C in RPMI 1640 medium supplemented with 10% heat-inactivated fetal bovine serum, penicillin (100 U/mL), streptomycin (100 µg/mL), and amphotericin B (0.25 µg/mL).
2. Primary fibroblasts are maintained on 100-mm dishes in 10 mL DMEM medium (high glucose, 4.5 g/L) supplemented with 20% fetal bovine serum, penicillin (100 U/mL), streptomycin (100 µg/mL), and amphotericin B (0.25 µg/mL), and are grown to 80–90% confluence prior to labeling. For metabolic labeling of fibroblasts, omit **Subheading 3.1., steps 1–6**, and proceed as follows: Wash fibroblasts (in one 100-mm dish) twice with 5 mL serum- and cysteine-free DMEM medium, and incubate in 4 mL of serum- and cysteine-free medium for 30 min. Add [^{35}S]-cysteine (100 µCi), and incubate overnight at 37°C (maximal incorporation of counts requires at least 12 h). Discard the medium, and wash the cells twice with 5 mL of ice-cold PBS, scrape the cells into 1.5 mL PBS, and pellet the cells by centrifugation in a benchtop centrifuge at 2400 rpm (500 g_{max} for 5 min at 4°C. Then proceed with extraction as described above (**Subheading 3.2.**).
3. Significant cell lysis can occur if cell pellets are pipeted too vigorously. Resuspend cells by gently tapping tubes.
4. For fibroblast-derived samples, more counts (30,000 cpm) should be loaded per lane, because more labeled material stays at the origin in preparations from fibroblasts as compared to lymphoblasts.
5. If desired, the lipid thioesters may be analyzed by HP-TLC in a second dimension for maximal resolution. The sample is applied to one corner of the plate and developed as described in **Subheading 3.3.** The plate is allowed to dry completely and a second tank is equilibrated in *n*-butanol:acetic acid:H_2O (60:20:20 [v/v/v]). The plate is turned 90° and developed in the second solvent system. For maximum reproducibility, the second run should be performed at a standard time interval after the first run, since the results are affected by residual solvent on the plate after the first run.

Acknowledgments

We thank Won Yi for technical assistance. This work was supported by Public Health Service Grant NS35323 from the National Institute for Neurological Disorders and Stroke.

References

1. Camp, L. A. and Hofmann, S. L. (1993) Purification and properties of a palmitoyl-protein thioesterase that cleaves palmitate from H-Ras. *J. Biol. Chem.* **268,** 22,566–22,574.
2. Camp, L. A., Verkruyse, L. A., Afendis, S. J., Slaughter, C. A., and Hofmann, S. L. (1994) Molecular cloning and expression of palmitoyl-protein thioesterase. *J. Biol. Chem.* **269,** 23,212–23,219.
3. Camp, L. A. and Hofmann, S. L. (1995) Assay and isolation of a palmitoyl-protein thioesterase from bovine brain using palmitoylated H-Ras as a substrate. *Methods Enzymol.* **250,** 336–347.
4. Verkruyse, L. A. and Hofmann, S. L. (1996) Lysosomal targeting of palmitoyl-protein thioesterase. *J. Biol. Chem.* **271,** 15,831–15,836.
5. Vesa, J., Hellsten, E., Verkruyse, L. A., Camp, L. A., Rapola, J., Santavuori, P., et al. (1995) Mutations in the palmitoyl protein thioesterase gene causing infantile neuronal ceroid lipofuscinosis. *Nature* **376,** 584–587.
6. Haltia, M., Rapola, J., Santavuori, P., and Keranen, A. (1973) Infantile type of so-called neuronal ceroid-lipofuscinosis. Part 2. Morphological and biochemical studies. *J. Neurol. Sci.* **18,** 269–285.
7. Haltia, M., Rapola, J., and Santavuori, P. (1973) Infantile type of so-called neuronal ceroid lipofuscinosis. Part I. Histological and electron microscopic studies. *Acta Neuropath.* **26,** 157–170.
8. Lu, J.-Y., Verkruyse, L. A., and Hofmann, S. L. (1996) Lipid thioesters derived from acylated proteins accumulate in infantile neuronal ceroid lipofuscinosis: Correction of the defect in lymphoblasts by recombinant palmitoyl-protein thioesterase. *Proc. Natl. Acad. Sci. USA* **93,** 10,046–10,050.
9. Soyombo, A. A. and Hofmann, S. L. (1997) Molecular cloning and expression of PPT2, a homolog of lysosomal palmitoyl-protein thioesterase with a distinct substrate specificity. *J. Biol. Chem.* **272,** 27,456–27,463.
10. Juguelin, H. and Cassagne, C. (1984) Assay of long-chain acyl-CoAs in a complex reaction mixture. *Anal. Biochem.* **142,** 329–335.

17

Fatty Acid Analysis of Protein-Derived Lipid Thioesters Isolated from Palmitoyl-Protein Thioesterase-Deficient Cells

Sandra L. Hofmann and Linda A. Verkruyse

1. Introduction

Cells from subjects with the neuronal storage disorder infantile neuronal ceroid lipofuscinosis (INCL) accumulate lipid thioesters derived from acylated proteins (*see* **ref. 1**) and Chapter 16). These lipid thioesters may play a role in the neurodegeneration caused by the deficiency in palmitoyl-protein thioesterase, the defective enzyme in this disorder *(2)*. In order to better characterize these thioesters, a method has been developed for their purification in quantities sufficient for characterization of the thioester fatty acid by high-performance liquid chromatography (HPLC). Lymphoblastoid cells (approx 1-L cultures) derived from patients with INCL are metabolically labeled with [^{35}S]cysteine and [^{3}H]palmitate (or other fatty acid, as desired), extracted into chloroform/methanol, and isolated by thin-layer chromatography (TLC) and HPLC. The highly purified lipids are hydrolyzed enzymatically using recombinant palmitoyl-protein thioesterase and the released fatty acids identified by migration on HPLC in comparison with known standards. An example is shown in **Fig. 1**.

In addition to palmitate, other radiolabeled fatty acids may be used for cell labeling, with the caveat that rapid metabolic conversion of radiolabeled palmitate to stearate (and potentially to other long-chain fatty acids) may occur. For example, stearic acid accounts for up to 40% of total ester-bound fatty acids in whole cellular proteins after labeling with [^{3}H]palmitate *(3,4)*. We have attempted to label the lipid thioesters with [^{3}H]acetate to achieve uniform labeling of fatty acids in proteins, as described by Towler and Glaser *(3)*, but insufficient counts were incorporated into the lipid thioesters to permit further analysis.

From: *Methods in Molecular Biology, Vol. 116: Protein Lipidation Protocols*
Edited by: M. H. Gelb © Humana Press Inc., Totowa, NJ

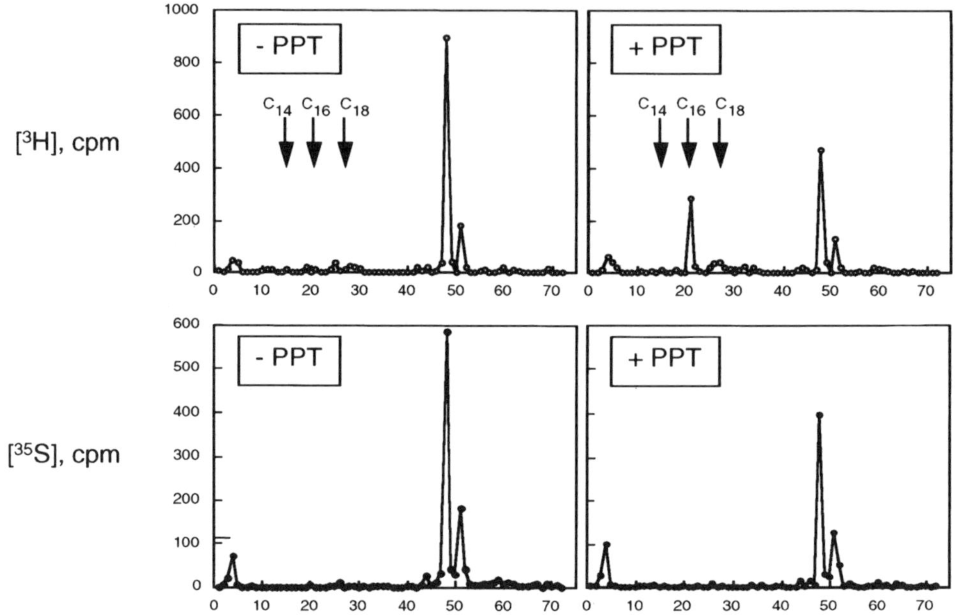

Fig. 1. Identification of palmitate as a fatty acid modifying one of the [^{35}S]cysteine-labeled lipid thioesters accumulating in INCL lymphoblasts. INCL lymphoblasts were metabolically labeled simultaneously with [^{3}H]palmitate and [^{35}S]cysteine and a chloroform/methanol extract was fractionated by TLC. One of the [^{35}S]cysteine-labeled bands was further purified by TLC and reverse and normal phase HPLC. The highly purified lipid thioester (shown here) was incubated with PPT (+PPT) or heat-inactivated PPT (–PPT), extracted into chloroform/methanol, and analyzed by C18 reverse-phase HPLC. The position of migration of myristate (C_{14}), palmitate (C_{16}), and stearate (C_{18}) standards is denoted by arrows (*upper panel*). The corresponding [^{35}S]cysteine counts, determined from the same samples shown in the upper panel, are shown in the *lower panel* for comparison.

2. Materials

2.1. Metabolic Labeling

1. RPMI 1640 (Gibco, Gaithersburg, MD, cat. no. 11875-01) tissue-culture medium supplemented with 10% heat-inactivated fetal bovine serum, penicillin (100 U/mL), streptomycin (100 µg/mL), and amphotericin B (0.25 µg/mL) (Gibco, cat. no. 15240-062).
2. [9,10-^{3}H]Palmitic acid (35.0 Ci/mmol) (NEN, Wilmington, DE) in ethanol.
3. [^{35}S]Cysteine (1231 Ci/mmol) (ICN, Costa Mesa, CA).
4. 15 mL Polystyrene tube.
5. Nitrogen gas (zero grade).

6. N-EVAP evaporator, nitrogen-delivery type (Organomation Associates, Berlin, MA).
7. Cysteine- and methionine-free RPMI 1640 tissue culture medium (Cellgro Mediatech, Herndon, VA, cat. no. 17-104-LI), to which supplemental methionine (Sigma, St. Louis, MO, cat. no. M2893) is added (final concentration, 15 μg/mL).
8. Incubator suitable for tissue culture set at 37°C.
9. Dulbecco's phosphate-buffered saline (PBS), without calcium or magnesium chloride (Gibco, cat. no. 14190-144).
10. Bench centrifuge.
11. Sodium pyruvate suitable for tissue culture.

2.2. Large-Scale Lipid Extraction

1. Chloroform/methanol 1:1 (v/v).
2. Chloroform/methanol 2:1 (v/v).
3. Two 50-mL glass centrifuge tubes.
4. Vortex mixer.
5. Bench centrifuge.
6. 10-mL glass pipets.
7. Dulbecco's PBS, without calcium or magnesium chloride (Gibco, cat. no. 14190-144).
8. Benzene.
9. Glass vials (20-mL liquid scintillation vials).
10. Nitrogen gas (zero grade).
11. N-EVAP evaporator, nitrogen-delivery type (Organomation Associates).

2.3. Thin-Layer Chromatography

1. LK6 silica gel 60 TLC plates, glass backed, with concentration zone, without fluorescent indicator, 20 × 20 cm, 0.25-mm thickness (Whatman, Clifton, NJ).
2. Chloroform, methanol, isopropanol.
3. Filter paper-lined chromatography tank with airtight lid.
4. Hamilton glass syringe for application of samples.
5. Fume hood.
6. Single-edged razor blades.
7. Nitrogen gas (zero grade).
8. N-EVAP evaporator, nitrogen-delivery type (Organomation Associates).
9. 10-mL glass pipets.
10. Cotton.
11. Sand.
12. Glass vials (20-mL liquid scintillation vials).
13. X-ray film, exposure cassette and fluorescent decals or marker.
14. Ring stand and clamps.

2.4. High-Performance Liquid Chromatography (HPLC)

1. HPLC system with programmable solvent module.
2. HPLC grade solvents: methanol, acetonitrile, isopropanol, hexane, methylene chloride, water.

3. Fraction collector.
4. 12 × 75 mm Glass tubes.
5. C4 semi-preparative reverse phase HPLC column (10 mm × 25 cm) (Vydac, Hesperia, CA, cat. no. 235341).
6. Normal phase silica HPLC column (4.6 mm × 25 cm) (Beckman, Houston, TX, Part No. 235341).
7. C18 reverse phase HPLC column (4.6 mm × 25 cm) (Beckman Part No. 235329).
8. 100-μL Hamilton glass syringe for sample injection.
9. Nitrogen gas (zero grade).
10. N-EVAP evaporator, nitrogen-delivery type (Organomation Associates).
11. Trifluoroacetic acid (TFA).
12. Benzene.
13. Glass vials (20-mL scintillation vials).
14. Liquid scintillation counter.
15. Scintillation fluid (#3a70B, Research Products International, Prospect, IL).
16. Myristic, palmitic, and stearic acids (5 mg/mL in ethanol).

2.5. Hydrolysis of Lipid Thioesters

1. Purified recombinant bovine palmitoyl-protein thioesterase.
2. Water bath set at 37°C.
3. 1.5 mL Microcentrifuge tubes.
4. Nitrogen gas (zero grade).
5. N-EVAP evaporator, nitrogen-delivery type (Organomation Associates).
6. Reaction buffer: 15 m*M* *n*-octylglucoside, 40 m*M* Tris-HCl, pH 7.0, 50 m*M* NaCl, 1 mM EDTA.
7. Vortex mixer.
8. Heated bead bath set at 100°C.

3. Methods

3.1. Metabolic Labeling with [^{35}S]Cysteine and [^{3}H]Palmitate

1. Add 12 mCi of [^{3}H]palmitate to a 15 mL polystyrene tube and evaporate the ethanol to approximately 20 μl under a stream of nitrogen (*see* **Note 1**).
2. Add 10 mL of prewarmed cysteine- and serum-free medium. Vortex vigorously.
3. Harvest 1.2 L of logarithmic-phase INCL B-lymphoblastoid cells (7.5 × 10^5 cells/mL) by centrifugation for 5 min at 220 g_{max}. Wash cells twice in prewarmed cysteine and serum free medium (*see* **Note 2**).
4. Resuspend the cell pellet in the [^{3}H]palmitate-containing medium, and adjust the cell density to 7.5 × 10^6 cells/mL with additional (approx 110 mL) cysteine- and serum-free medium. Adjust the labeling medium to a final concentration of 5 m*M* pyruvate and add 6 mCi of [^{35}S]cysteine.
5. Incubate cells for 5 h at 37°C.
6. Harvest labeled cells by centrifugation at 220 g_{max} for 5 min at 4°C. Wash cells twice in ice-cold PBS and pellet cells by centrifugation at 220 g_{max} for 5 min at 4°C.

3.2. Large-Scale Extraction of Labeled Lipid Thioesters

1. Resuspend the labeled cell pellet in 5 mL of PBS (200 μL PBS/3.5 × 10^7 cells).
2. Mix the resuspended cell pellet with 30 mL of chloroform/methanol 1:1 (v/v) in a 50-mL glass centrifuge tube. Vortex the sample intermittently for 30 min at room temperature.
3. Centrifuge the sample for 5 min at 400 g_{max} to separate the phases. Transfer the organic (lower) phase to a clean 50-mL glass centrifuge tube.
4. Add 6 volumes chloroform/methanol 2:1 (v/v) to the aqueous (upper) phase and extract a second time.
5. Wash the combined organic phases with one-sixth volume cold PBS. Transfer the organic (lower) phase to a glass vial, dilute samples with one-half volume benzene, and evaporate under a stream of nitrogen.
6. Dissolve the residue in 2–3 mL of 1:1 chloroform/methanol.

3.3. Thin-Layer Chromatography of Lipid Thioesters

1. Add solvent (chloroform/methanol/water 65:25:4 [v/v/v]) to a chromatography tank lined with absorbent paper to a depth of 0.5–1.0 cm. Seal the lid on the tank and allow the system to equilibrate for 2 h (*see* **Note 3**).
2. Apply extracted lipids to eight LK6 silica gel 60 plates (20 × 20 cm, 0.25 mm thickness). Use the entire width of the plate (*see* **Note 4**). To facilitate identification of the [^{35}S]-labeled bands, *see* **Note 5**.
3. Develop plates in the tank prepared in **step 1** until the solvent travels 17 cm.
4. Air-dry plates for 15–30 min in a fume hood at room temperature.
5. Mark plates with a fluorescent decal or radioactive pen, apply plates directly to film, and expose overnight at –85°C.
6. Place the developed film over the plates and align marks. Outline the position of the labeled band(s) with pencil. Scrape the marked area(s) from the plates with a single edged razor blade. Load the silica into cotton-plugged 10-mL glass pipets. Cover the silica with a 3 mm layer of sand.
7. Elute the lipids from the silica with 30 mL chloroform/isopropanol 1:1 (v/v) followed by 30 mL chloroform/methanol 1:1 (v/v) into glass vials. Combine the two fractions.
8. Evaporate the solvents in a fume hood under a nitrogen stream, or use a rotary evaporator, if available (*see* **Note 6**). Redissolve lipids in 2 mL chloroform/methanol 1:1 (v/v).

3.4. High-Performance Liquid Chromatography

3.4.1. C4 Reverse-Phase Semi-Preparative HPLC

1. Concentrate lipids to 100 μL by evaporation under a nitrogen stream (*see* **Note 7**).
2. Prepare solvents for HPLC separation. Solvent system 1:0.1% TFA in water, solvent system 2:0.1% TFA in acetonitrile/isopropanol 1:1 (v/v).
3. Program HPLC solvent gradient as follows. (This programming results in isocratic elution at 50% solvent 2 for 5 min, a gradient between 50 and 90% solvent 2 for 20 min, isocratic elution at 90% solvent 2 for 10 min, a gradient

from 90–100% solvent 2 for 10 min, isocratic elution at 100% solvent 2 for 15 min, a gradient from 100%–50% solvent 2 for 5 min, and isocratic elution at 50% solvent 2 for 5 min.)

Time (min)	%Solvent 2
0	50
5	50
25	90
35	90
45	100
60	100
65	50
70	50

4. Load the sample onto the column at a flow rate of 2.0 mL/min. Collect 2.0 mL fractions.
5. Evaporate solvents from 100 µL of each fraction for liquid scintillation counting of [^{3}H] and [^{35}S].
6. Pool fractions containing peak [^{35}S] activity. Dilute with an equal volume of benzene and evaporate under a nitrogen stream. Re-dissolve in 1.5 mL chloroform/methanol 1:1 (v/v).

3.4.2. Normal Phase HPLC

1. Concentrate the sample to 100 µL by evaporation under a nitrogen stream.
2. Prepare solvents for HPLC. Solvent system 1: Hexane; solvent system 2: methylene chloride/methanol 2:3 (v/v).
3. Program HPLC solvent gradient as follows:

Time (min)	% Solvent 2
0	0
5	0
55	100
60	100
65	0
70	0

4. Load the sample onto the column at a flow rate of 1.0 mL/min. Collect 1.0 mL fractions.
5. Evaporate solvents from 50 µL of each fraction for liquid scintillation counting of [^{3}H] and [^{35}S].
6. Pool fractions containing peak [^{35}S] activity. Store samples at –20°C for treatment with palmitoyl-protein thioesterase (PPT) enzyme.

3.5. Large-Scale Enzymatic Hydrolysis and Fatty Acid Analysis

1. Add lipids containing from 1700–2000 cpm in [^{35}S] to two microfuge tubes and evaporate the solvents under a stream of nitrogen (*see* **Note 8**).

2. Add 95 μl of reaction buffer (15 m*M* *n*-octylglucoside, 40 m*M* Tris-HCl, pH 7.0, 50 m*M* NaCL, 1 m*M* EDTA) to each tube and vortex samples vigorously for 60 s.
3. Heat inactivate 2.5 μg of recombinant PPT in a bead bath at 100°C for 15 min.
4. To one tube from **step 2** add 2.5 μg PPT, and to the second tube add the heat-inactivated PPT.
5. Incubate samples at 37°C for 90 min. Stop reaction by placing tubes on ice. Count a 5 μL aliquot of each tube to verify solubilization of radioactivity.
6. Prepare solvents for HPLC separation. Solvent system 1: 0.1% TFA in water, solvent system 2: 0.1% TFA in acetonitrile/isopropanol 1:1 (v/v).
7. Program HPLC solvent gradient as follows:

Time (min)	% Solvent 2
0	75
5	75
20	90
30	90
45	100
65	100
70	75
75	75

8. Load samples (100 μL) onto the column at a flow rate of 1.0 mL/min. Collect 1.0 mL fractions.
9. Evaporate solvent from each fraction and count by liquid scintillation for [^{3}H] and [^{35}S].
10. Inject 100-μL aliquots of fatty acid standards in separate runs and monitor elution time by absorbance at 220 nm.

4. Notes

1. Solubilization of the labeled palmitate is significantly improved if a small amount of the evaporating ethanol is retained prior to addition of the labeling medium. It may be prudent to verify complete solubilization by liquid scintillation counting before adding the labeling medium to the cells.
2. Significant cell loss can occur if cell pellets are pipeted vigorously. Resuspend cells by gently tapping tubes.
3. For consistent TLC results, solvents should be prepared fresh daily.
4. Preparative TLC plates may be used to reduce the number of plates needed, but some loss of resolution may be expected.
5. During TLC, it is useful to apply a crude lipid extract from cells labeled with [^{35}S]cysteine only (10,000 cpm) to a 0.5-cm lane at one end of the TLC plates to facilitate identification of the [^{35}S]-labeled compounds, as there may be some spillover of [^{3}H] counts onto the autoradiographic film.
6. To minimize sample loss through surface binding to containers, concentrate samples in glass tubes containing a pointed bottom. Wash sides with solvent

repeatedly, or concentrate only a portion, add a second aliquot, and repeat until the desired volume is reached.

7. Prior to loading samples onto HPLC columns, it may be desirable to centrifuge for 1 min at maximum speed to pellet any particulate material that may disrupt column flow.
8. Fatty acids may also be released from the thioesters by chemical methods, such as incubation in the presence of neutral hydroxylamine, followed by conversion of the fatty acid hydroxamates to fatty acid methyl esters *(3)*.

Acknowledgments

We thank Won Yi for technical assistance. This work was supported by Public Health Service Grant NS35323 from the National Institute for Neurological Disorders and Stroke and the Medical Scientist Training Program.

References

1. Lu, J.-Y., Verkruyse, L. A., and Hofmann, S. L. (1996) Lipid thioesters derived from acylated proteins accumulate in infantile neuronal ceroid lipofuscinosis: correction of the defect in lymphoblasts by recombinant palmitoyl-protein thioesterase. *Proc. Natl. Acad. Sci. USA* **93,** 10,046–10,050.
2. Vesa, J., Hellsten, E., Verkruyse, L. A., Camp, L. A., Rapola, J., Santavuori, P., Hofmann, S. L., and Peltonen, L. (1995) Mutations in the palmitoyl protein thioesterase gene causing infantile neuronal ceroid lipofuscinosis. *Nature* **376,** 584–587.
3. Towler, D. and Glaser, L. (1986) Acylation of cellular proteins with endogenously synthesized fatty acids. *Biochemistry* **25,** 878–884.
4. McIlhinney, R. A. J., Chadwick, J. K., and Pelly, S. H. (1987) Studies on the cellular location, physical properties and endogenously attached lipids of acylated proteins in human squamous-carcinoma cell lines. *Biochem. J.* **244,** 109–115.

Index